Judgment Aggregation: A Primer

Synthesis Lectures on Artificial Intelligence and Machine Learning

Editors
Ronald J. Brachman, *Yahoo! Labs*
William W. Cohen, *Carnegie Mellon University*
Peter Stone, *University of Texas at Austin*

Markov Logic: An Interface Layer for Artificial Intelligence
Pedro Domingos and Daniel Lowd
2009

Introduction to Semi-Supervised Learning
XiaojinZhu and Andrew B.Goldberg
2009

Action Programming Languages
Michael Thielscher
2008

Representation Discovery using Harmonic Analysis
Sridhar Mahadevan
2008

Essentials of Game Theory: A Concise Multidisciplinary Introduction
Kevin Leyton-Brown and Yoav Shoham
2008

A Concise Introduction to Multiagent Systems and Distributed Artificial Intelligence
Nikos Vlassis
2007

Intelligent Autonomous Robotics: A Robot Soccer Case Study
Peter Stone
2007

Judgment Aggregation: A Primer
Davide Grossi and Gabriella Pigozzi

ISBN: 978-3-031-00440-7   paperback
ISBN: 978-3-031-01568-7   ebook

DOI 10.1007/978-3-031-01568-7

A Publication in the Springer series
SYNTHESIS LECTURES ON ARTIFICIAL INTELLIGENCE AND MACHINE LEARNING

Lecture #27
Series Editors: Ronald J. Brachman, *Yahoo! Labs*
             William W. Cohen, *Carnegie Mellon University*
             Peter Stone, *University of Texas at Austin*
Series ISSN
Synthesis Lectures on Artificial Intelligence and Machine Learning
Print 1939-4608   Electronic 1939-4616

Judgment Aggregation: A Primer

Davide Grossi
University of Liverpool

Gabriella Pigozzi
Université Paris Dauphine

SYNTHESIS LECTURES ON ARTIFICIAL INTELLIGENCE AND MACHINE LEARNING #27

ABSTRACT

Judgment aggregation is a mathematical theory of collective decision-making. It concerns the methods whereby individual opinions about logically interconnected issues of interest can, or cannot, be aggregated into one collective stance. Aggregation problems have traditionally been of interest for disciplines like economics and the political sciences, as well as philosophy, where judgment aggregation itself originates from, but have recently captured the attention of disciplines like computer science, artificial intelligence and multi-agent systems. Judgment aggregation has emerged in the last decade as a unifying paradigm for the formalization and understanding of aggregation problems. Still, no comprehensive presentation of the theory is available to date. This Synthesis Lecture aims at filling this gap presenting the key motivations, results, abstractions and techniques underpinning it.

KEYWORDS

judgment aggregation, collective decision-making, logic, social choice theory, computational social choice, preference aggregation, voting paradoxes, aggregation rules, impossibility results, manipulability, ultrafilters, opinion pooling, deliberation

DG—to my grandmother Alice, in memoriam

GP—to my mother and to the memory of my father

Contents

Preface

This book concerns the aggregation of individual opinions into group opinions. When opinions exhibit logical structure (e.g., accepting that p is the case and accepting that p *implies* q compels me to also accept that q is the case) aggregation becomes difficult. Is it possible at all to find aggregation procedures that preserve compliance with logical principles, and that at the same time appeal to democratic criteria like, for instance, not being dictatorial? Are the methods we commonly use to aggregate our opinions (e.g., majority voting) appropriate, and under which conditions? And if, after all, ideal procedures turn out to be impossible, what are the reasons for such impossibility? Questions like these are the playground of judgment aggregation, and will be the topic of this book.

Before starting, the reader can find here some information about the main objectives we pursued by writing the book, the readership we aimed at, and an outline of the topics we are going to cover.

Objectives   In writing this introductory book on judgment aggregation we had two main objectives in mind. First, we wanted to provide a compact and systematic exposition of the problems, definitions, results and proof techniques that drive the field. Survey papers appeared in philosophical and social sciences journals and volumes [LP09, Car11, Mon11, Lis12], but no comprehensive exposition of the field is available to date. Second, we wanted to make the theory of judgment aggregation accessible, in a 'sympathetic' format, to the disciplines of artificial intelligence and multi-agent systems, which in recent years have increasingly been concerned with the problems of aggregation and voting.

Readership and prerequisites   The book is primarily meant as an introduction to the field of judgment aggregation for graduate students and researchers in computer science, artificial intelligence and multi-agent systems. At the same time, it has been our aim to make the book accessible also to mathematically minded graduate students and researchers in philosophy, the social and the political sciences. The material is presented in such a way to presuppose only familiarity with propositional logic and basic discrete mathematics. The book intends to put the reader at pace with the field, enabling the key conceptual, technical and bibliographical tools to understand (and possibly contribute to) its current developments. We have not included any exercises, but the reader will be asked at times to complete missing steps in proofs or try to prove statements given as running comments in the main text.

Outline of the book   The book is structured in two parts. The first part (Chapters 1–4) introduces what can be considered the established body of the theory: the motivating examples behind judgment aggregation and its place within the field of social choice theory (Chapter 1); the logic-based framework for judgment aggregation (Chapter 2); some results, with proofs, on the impossibility of finding 'ideal' aggregation procedures (Chapter 3); and various ways that have been explored in the literature to work around the limits imposed by those results (Chapter 4). The second part (Chapters 5–7) touches upon topics that are, to a greater or lesser extent, still under development in the research agenda of the field: the issue of manipulation and strategic behavior in judgment aggregation (Chapter 5); the design of non-resolute aggregation procedures (Chapter 6); and the modeling of deliberative processes of aggregation, and of processes of pre-vote deliberation (Chapter 7). All chapters present an overview of key concepts and results and conclude with a section pointing to further topics and readings and, sometimes, open issues.

Davide Grossi and Gabriella Pigozzi
December 2013

Acknowledgments

The present book has grown out of materials developed for a course entitled "Introduction to Judgment Aggregation" taught at *ESSLLI'11* (Ljubljana, Slovenia) and considerably extends in both depth and coverage the lecture notes we wrote for that course, published in [GP12]. Parts of the book have also been presented at the *Stanford Logic Group Seminar* of Stanford University (Spring 2011), and at the *Computer Science Seminar* of the Royal Holloway University of London (Autumn 2013). We are much indebted to the students who attended the *ESSLLI'11* course and the participants of the above events for their questions, comments and suggestions. They provided essential feedback for the writing of the book.

We are greatly indebted to Denis Bouyssou, Umberto Grandi, Eric Pacuit and the anonymous reviewers arranged by the publisher, who thoroughly read earlier drafts of the book and gave us detailed and extremely helpful feedback to produce this final version. The book largely benefited also from discussions with Franz Dietrich, Ulle Endriss and Umberto Grandi. Any remaining errors and omissions are of course the sole responsibility of the authors. Finally, we would like to thank Michael Wooldridge for encouraging us to embark in this project, and Morgan & Claypool—in particular our editor Mike Morgan—for assisting us along the way and making the writing of this book such a pleasant journey.

Davide Grossi and Gabriella Pigozzi
December 2013

CHAPTER 1

Logic Meets Social Choice Theory

Judgment aggregation is a recent theory that combines aggregation problems previously studied by social choice theory with logic. Social choice theory is a vast subject, including not only the study of preference aggregation and voting theory but also topics like social welfare and justice. Given the tight links between judgment aggregation and preference aggregation, in this first chapter we give a concise survey on some historical aspects of preference aggregation and then introduce and motivate the more recent field of judgment aggregation. The chapter builds on [Sen99, Sen86, Bla58] for the historical overview on social choice theory, and on [KS93, Kor92] for the informal introduction to judgment aggregation.

Chapter outline: We start by giving a brief overview of the history of social choice theory, from the contributions of Borda and Condorcet during the French Revolution (Section 1.1.1) until the general impossibility theorem by Arrow in 1951 that started modern social choice theory (Section 1.1.2). We then present the doctrinal paradox that originated the whole field of judgment aggregation (Section 1.2), and look at how judgment aggregation relates to the older theory of preference aggregation (Section 1.2.2). In the concluding section we point to literature at the interface of social choice theory, computer science and artificial intelligence, completing the sketch of the broad scientific context of the present book.

1.1 A CONCISE HISTORY OF SOCIAL CHOICE THEORY

Social choice theory studies how individual preferences and interests can be combined into a collective decision. An example of such a type of aggregation problems is a group electing one of many candidates on the basis of the preferences that the individuals in the group express over the candidates.

Collective decision-making is a constant feature of human societies. In his 1998 Nobel lecture, Amartya Sen [Sen99] recalled that already in the fourth century B.C., Aristotle in Greece and Kautilya in India explored how different individuals could take social decisions. However, the systematic and formal study of voting and committee decisions started only during the French Revolution, thanks to French mathematicians like Borda, Condorcet and Laplace.

Besides the problem of selecting the winning candidate in an election, social choice theory has its origins in the normative analysis of welfare economics [Sen86], a branch of modern economics that evaluates economic policies in terms of their effects on the social welfare of the members of the society. Welfare economics took inspiration from Jeremy Bentham's utilitarianism rather than from Borda and Condorcet. According to Bentham's utilitarianism fundamental axiom, "it is the greatest happiness of the greatest number that is the measure of right and wrong" [Ben76, Preface (2nd para.)]. Utilitarianisms assumed that individual preferences could be expressed by cardinal utilities and that they could then be compared across different individuals (interpersonally comparable preferences). Yet, in the 1930s both cardinality and interpersonal comparability of personal utility were questioned by Lionel Robbins [Rob38]. If utilities reflect individual mental states—Robbins argued—since it is not possible to measure mental states, then utilities cannot be compared across several individuals either. It was such "informational restriction" [Sen86] of social welfare to a n-tuple of ordinal (and interpersonally non-comparable) individual utilities that induced economists to look at methods developed in the theory of elections. As we shall see in Section 1.1.2, the informational restriction that followed Robbins's criticisms made the problem of welfare economics look similar to the exercise of deriving a social preference ordering from individual orderings of social states, a problem addressed by Borda and Condorcet during the French Revolution.

1.1.1   THE EARLY HISTORY

On the history of the theory of elections McLean wrote:

> The theory of voting is known to have been discovered three times and lost twice. The work of Condorcet, Borda, and Laplace was entirely ignored from about 1820 until 1952, with the sole exception of E. J. Nanson's paper 'Methods of Election', which was read to the Royal Society of Victoria in 1882, published in a British Government Blue Book of 1907, and languished there undiscussed until 1958. C. L. Dodgson ('Lewis Carroll') discussed Condorcet and Borda methods, and procedures for breaking cycles, in three pamphlets printed in the 1870s; he worked in ignorance of his predecessors, and again was not understood until 1958. [McL90, p. 99]

However, there have been precedents to the work of those scholars. In particular, McLean [McL90] discovered that a method developed by Condorcet was proposed as early as in the thirteenth century by Ramon Lull, and that a method developed by Borda was introduced in the fifteenth century by Nicolas Cusanus. So, what are these Condorcet and Borda methods and why are they so important in the history of social choice?

Borda

Jean-Charles de Borda, a French mathematician member of the Academy of Sciences, developed the first mathematical theory of elections. According to Black,[1] Borda read the paper before the

---

[1]Duncan Black (called the founder of social choice [Tul91] for being the first to really understand the work of Condorcet and discovering Dodgson's papers) gave in [Bla58] an excellent historical overview of the mathematical theory of voting starting from Borda, Condorcet and Laplace and including Nanson, Galton and Dodgson.

25	x	y	z
20	y	z	x
15	z	y	x

Figure 1.1: A problem with plurality voting.

Academy of Sciences already in 1770. However, the report that was supposed to be written about Borda's essay was never accomplished. Fourteen years later, a report on a manuscript by Marie Jean Antoine Nicolas Caritat (better known as the Marquis de Condorcet) was presented at the Academy. Few days later, Borda read for the second time his paper, which was printed in 1781, but published only in 1784 [Bor84]. Borda method was adopted by the Academy as the method to elect its members. It was used until 1800 when a new member, Napoleon Bonaparte, attacked it.

In his *Mémoire sur les élections au scrutin*, Borda first showed that *plurality voting*, probably the most well-known voting method, is not satisfactory as it may elect the wrong candidate. In plurality voting, each individual votes one candidate, and the candidate that receives the greatest number of votes is elected. The problem with this procedure is that it ignores the individual preferences over candidates. Suppose, for example, that there are three alternatives x, y and z and 60 voters. Of these 60 voters, 25 prefer x to y and y to z, 20 prefer y to z and z to x and, finally, 15 prefer z to y and y to x, as shown in Figure 1.1, where preferences are given in a left to right order.

Assuming that the individuals vote for the candidate at the top of their preferences, we obtain that x gets 25 votes, y gets 20 votes and z only 15. Thus, if plurality vote is used, x will be selected. However, Borda noticed that for a majority of the voters, x is the least preferred candidate: pairwise majority comparison shows that 35 voters against 25 would prefer both y and z to x. Plurality vote selects the candidate that receives the most votes but not necessarily more than half of the votes in pairwise comparisons. Thus, the two procedures (plurality and pairwise majority) can lead to different outcomes. What is interesting is that, as observed by Black, in his argument Borda really made use of what is now known as the *Condorcet criterion*, according to which a voting system should select the candidate that defeats every other candidate. When it exists, such a candidate is unique and is called the *Condorcet winner*. However, Borda did not develop this line of thought. We have to await Condorcet for such a principle to be clearly put forward.

The solution proposed by Borda to the fact that plurality may select the wrong candidate is a method which makes use of the entire order in the voters' preferences. In his method, voters rank all the candidates (assumed to be finite). If there are n candidates, each top place candidate gets n points, each candidate at the second place gets $n - 1$ points, and so on until the least preferred candidate, which gets 1 point. The alternative with the highest total score is elected. Borda's rank-order method is an example of what we would call today a *scoring rule*.[2] Scoring rules are a class of standard aggregation rules in preference aggregation [You74, You75].

Let us suppose that a voter prefers x to y and y to z. The Borda method rests on two assumptions. The first is the measurability of utility, i.e. (paraphrasing Borda) that the degree of superiority that the voter gives to x over y should be considered the same as the degree of superiority that he gives to y over z. The second is interpersonal utility, that is, how different individual utilities can be measured. In Borda method, voters are given equal weight. The justification that Borda provides for the first assumption is based on ignorance: there is no reason to assume that, by placing y between x and z, the voter wanted to place y nearer to x than to z. The second is justified on the basis of equality among voters. At the end of his paper, he claims that his method can be used in any kind of committee decision. Even though Borda fails to thoroughly examine the nature of collective decisions [Bla58], he realized that his method was open to manipulation, that is, to the possibility of voters misrepresenting their true preferences to the rule in order to elect a better (according to their true preferences) candidate.[3] In particular, a voter could place the strongest competitors to his most preferred candidate at the end of the ranking. Addressing this issue Borda famously replied: "My scheme is only intended for honest men."

Condorcet

The other famous member of the Academy of Sciences was Condorcet. His work on the theory of elections is mostly contained in the mathematical (and hardly readable) *Essai sur l'Application de l'Analyse à la Probabilité des Décisions Rendues à la Pluralité des Voix* [Con85]. Borda and Condorcet were friends and in a footnote in his *Essai*, Condorcet says that he completed his work before he was acquainted with Borda's method.

As Black traced back, there are really two approaches in Condorcet's work. The first contribution is in line with Borda. Like Borda, Condorcet observes that plurality vote may result in the election of a candidate against which each of the other candidates has a majority. This led to the formulation of the above mentioned Condorcet criterion, that is, the candidate to be elected is the one that receives a majority against each other candidate (instead of just the highest number of votes). Whereas Borda employed a positional approach, Condorcet recommended a method based on the pairwise comparison of alternatives. Given a set of individual preferences, the method suggested by Condorcet consisted in the comparison of each of the alternatives in pairs. For each pair, the winner is determined by majority voting, and the final collective ordering is obtained by a combination of all partial results. The *Condorcet winner* is the candidate that beats every other alternative in a pairwise majority comparison. However, he also discovered a disturbing problem of majority voting, now known as the *Condorcet paradox*. He discovered that pairwise majority comparison may lead a group to hold an intransitive preference (or a *cycle*, as later called by Dodgson) of the type that x is preferred to y, y is preferred to z, and z to x. This is the cycle

---

[2]The same method was also suggested by Laplace in 1795, in a series of lectures he gave at the École Normale Supériéure in Paris.

[3]Manipulation will be the topic of Chapter 5.

$$\begin{array}{c|ccc} \text{Voter 1} & x & y & z \\ \text{Voter 2} & y & z & x \\ \text{Voter 3} & z & x & y \end{array}$$

Figure 1.2: An illustration of the Condorcet paradox.

we obtain if we consider, for example, three voters expressing preferences as in Figure 1.2, where preferences are given in a left to right order and voter 1 prefers x to y and y to z, voter 2 prefers y to z and z to x, while voter 3 prefers z to x and x to y.

The trouble with a majority cycle is that the group seems unable to single out the 'best' alternative in a principled manner. Note also that devising rules fixing some order in which the alternatives are to be compared does not solve the problem. For instance, if in the example above we fix a rule that compares alternatives x and y first and the winner is then pitted against z, alternative z would win the election. However, if we instead choose to compare x and z first and then to compare the winning alternative with y, we would get a different result, namely y would be the winning alternative.

Condorcet's second main contribution employs probability theory to deal with the 'jury problem'. Voters are seen like jurors who vote for the 'correct' alternative (or the 'best' candidate). The idea that groups make better decisions than individuals dates back to Rousseau [Rou62], according to whom, in voting, individuals express their opinions about the 'best' policy, rather than personal interests. Condorcet approached Rousseau's position in probabilistic terms and aimed at an aggregation procedure that would maximize the probability that a group of people take the right decision. This led Condorcet to formulate the result now known as the *Condorcet Jury Theorem*, which provides an epistemic justification to majority rule [GOF83]. The theorem states that, when all jurors are independent and have a probability of being right on the matter at issue, which is higher than random, then majority voting is a good truth-tracking method. In other words, under certain conditions, groups make better decisions than individuals, and the probability of the group taking the right decision approaches 1 as the group size increases.

So, interestingly, Condorcet showed at the same time the possibility of majority cycles, a negative result around which much of the literature on social choice theory built up, and a positive result like the Condorcet Jury Theorem, which gives an epistemic justification to majority voting.[4]

Dodgson
From the overview so far, the reader may have gotten the impression that the early developments of social choice theory were exclusively due to French scholars. But this is not the case. Indeed

---

[4]It is worth observing, in passing, that these two landmark results can be viewed as stemming from two different ways of conceiving democracy: the first one sees democracy as based on *preferences*, while the second sees it as based on *knowledge* (epistemic conception, as called in [Coh86, CF86]).

many English mathematicians have also studied the subject: Eduard John Nanson, Francis Galton and, more importantly, the Rev. Charles Lutwidge Dodgson (better known as 'Lewis Carroll', author of *Alice's Adventures in Wonderland*), to whom we now turn.

Black gives a careful analysis of Dodgson's life and of the circumstances that raised the interests of a Mathematics lecturer at Christ Church college in Oxford for the theory of elections. In particular, Black discovered three of Dodgson's previously unpublished pamphlets and, thanks to his extensive research, could conclude that Dodgson ignored the works of both Borda and Condorcet.

Dodgson referred to well-known methods of voting (like plurality and Borda's method) and highlighted their deficiencies. For him the main interest of the theory of elections resided in the existence of majority cycles. He suggested a modification of Borda's method to the effect of introducing a 'no election' alternative among the existing ones [Dod73], the idea being that in case of cycles, the outcome should be 'no election'. He then claimed that if there is no Condorcet winner, his modified method of marks should be used [Dod74].

Later Dodgson proposed a method based on pairwise comparison that may seem to contradict the 'no election' principle he introduced earlier. However, as Arrow suggests [Arr63], this approach could be used when we do not wish to accept 'no election' as a possible outcome. The new method (now known as *Dodgson rule*) selects the Condorcet winner (whenever there is one) and otherwise finds the candidate that is 'closest' to being a Condorcet winner [Dod76]. The idea is to find the (not necessarily unique) alternative that can be made a Condorcet winner by a minimum number of preference switchings in the original voters' preferences. A switch is a preference reversal between two adjacent positions. In order to illustrate the method, let us consider one of the examples made by Dodgson himself.

Consider the preference profile given in Figure 1.3. Each row represents a group of voters with the same preferences, given in a left to right order. The number in the first column indicates the size of each group. In this example, there are eleven voters and four alternatives ($a, b, c,$ and d). As the reader can easily check, the majority is cyclical ($adcba$) and none of the alternatives is a Condorcet winner. However, if the voter holding preference $dcba$ switches alternatives c and b (marked by an asterisk) in her preference ranking, b becomes a Condorcet winner. Alternative c also can be made a Condorcet winner by one switch, so b and c are the only Dodgson winners (a and d each need four switches to be preferred to every other alternative by some strict majority).

1.1.2 MODERN SOCIAL CHOICE THEORY

We have mentioned how Robbins's claim [Rob38] that interpersonal utilities could not be compared undermined what constituted the predominant utilitarian approach to welfare economics until the Thirties: this amounted to say that there is social improvement when everyone's utility goes up (or, at least, no one's utility goes down when someone's utility goes up) [Sen95].

It thus appeared that social welfare must be based on just the n-tuple of ordinal interpersonally non-comparable, individual utilities. [...] This "informational crisis" is

2	a	d	c	b
2	a	b	d	c
2	b	c	a	d
1	b	d	c	a
3	c	b	d	a
1	d	$c*$	$b*$	a

Figure 1.3: An example of Dodgson's rule.

important to bear in mind in understanding the form that the origin of modern social choice theory took. In fact, with the binary relation of preference replacing the utility function as the primitive of consumer theory, it made sense to characterise the exercise as one of deriving a social preference ordering R from the n-tuple of individual orderings $\{R_i\}$ of social states. [Sen86, p. 1074]

The need for functions of social welfare defined over all the alternative social states was made explicit by Abram Bergson [Ber38, Ber66] and Paul Samuelson [Sam47]. Economists turned to the mathematical approach to elections explored by Condorcet, Borda, and Dodgson only when—following the informational restriction decreed by Robbins—they searched for methods to aggregate binary relations of preference into a social preference ordering. Thus, social choice theory stemmed from two distinct problems—how to select the winning candidate in an election, and how to define social welfare—and the relations between these problems became clear only in the 1950s.

Young economist and future Nobel prize winner, Kenneth Arrow defined a *social welfare function* as a function that maps any n-tuple of individual preference orders to a collective preference order. His axiomatic method outlined the requirements that any desirable social welfare function should satisfy.[5] In 1950 he proved what still is the major result of social choice, the "General Possibility Theorem," now better known as *Arrow's impossibly result*[6] [Arr50, Arr63]. The theorem shows that there exists no social welfare function that satisfies only just a small number of desirable conditions.

Let us informally present these conditions: the first is that a social welfare function must have a *universal domain*, that is, it has to accept as input any combination of individual preference orders. Another commonly accepted requirement is the *Pareto condition*, which states that, whenever all members of a society rank alternative x above alternative y, then the society must also prefer x to y. The *independence of irrelevant alternatives* condition states that the social preference

[5]See [Sup05] for a reconstruction of the intellectual path that led Arrow to introduce the axiomatic method in economics, and in particular Alfred Tarski's influence, of whom Arrow attended a course in the calculus of relations as an undergraduate student.

[6]In Chapter 4 we will come back to the subtle relationships between impossibility and possibility theorems.

over any two alternatives x and y must depend only on the individual preferences over those alternatives x and y (and not on other—irrelevant—alternatives).[7] Finally, *non-dictatorship* requires that there exists no individual in the society such that, for any domain of the social welfare function, the collective preference is the same as that individual's preference (i.e., the dictator). Arrow's celebrated result shows that no social welfare function can jointly satisfy these conditions.[8]

1.2   A NEW TYPE OF AGGREGATION

1.2.1   FROM THE DOCTRINAL PARADOX TO THE DISCURSIVE DILEMMA

We have seen that, thanks to the Condorcet Jury Theorem, majority rule enjoys an attractive property: some conditions being satisfied, groups make better decisions than individuals. Yet, unfortunately, the Condorcet paradox also showed that this same rule is unable to ensure consistent social positions under all situations.

Classical social choice theoretic models focus on the aggregation of individual preferences into collective outcomes. Such models focus primarily on collective choices between alternative outcomes such as candidates, policies or actions. However, they do not capture decision problems in which a group has to form collectively endorsed beliefs or judgments on logically interconnected propositions. Such decision problems arise, for example, in expert panels, assemblies and decision-making bodies as well as artificial agents and distributed processes, seeking to aggregate diverse individual beliefs, judgments or viewpoints into a coherent collective opinion. Judgment aggregation fills this gap by extending earlier approaches developed by social choice theory for the aggregation of preferences.[9]

Doctrinal paradox

Judgment aggregation has its roots in jurisprudence. The paradox of a group of rational individuals collapsing into collective inconsistency made its first appearance in the legal literature, where constitutional courts are expected to provide reasons for their decisions. The discovery of the paradox was attributed to Kornhauser and Sager's 1986 paper [KS86]. However, Elster recently pointed out that structurally similar problems have been first indicated by Poisson in 1837 [Els13]. What is now known as the *doctrinal paradox* [KS93, Kor92, Cha98] was rediscovered in 1921 by the Italian legal theorist Vacca [Vac21] (see [Spe09]), who consequently raised severe criticisms to the possibility of deriving collective judgments from individual opinions. The logical problem of aggregation was also noticed by Guilbaud [Gui52, Mon05], who gave a logical interpretation to preference aggregation.

---

[7]This, as we shall see, is a more controversial requirement.

[8]It is impossible to underestimate the influence that Arrow's theorem had in the development and foundation of social choice as a formal discipline. His result generated a vast literature, including many other impossibility results, like [Bla57, Sen69, Sen70, Pat71, Gib73, Sat75], to quote only few of them. Political scientists (most notably, William Riker [Rik82]) argued that Arrow's findings posed serious threat to the theory of democracy.

[9]On the relations between judgment aggregation and preference aggregation, see Sections 1.2.2 and 3.4.

	Valid contract p	Breach q	Defendant liable r
Judge 1	1	1	1
Judge 2	1	0	0
Judge 3	0	1	0
Majority	1	1	0

Figure 1.4: An illustration of the doctrinal paradox.

In order to illustrate the doctrinal paradox, we recall the familiar example in the literature by Kornhauser and Sager [KS93]. A three-member court has to reach a verdict in a breach of contract case between a plaintiff and a defendant. According to the contract law, the defendant is liable (the *conclusion*, here denoted by proposition r) if and only if there was a valid contract and the defendant was in breach of it (the two *premises*, here denoted by propositions p and q respectively). Suppose that the three judges cast their votes as in Figure 1.4.

The court can rule on the case either directly, by taking the majority vote on the conclusion r regardless of how the judges voted on the premises (*conclusion-based procedure*) or indirectly, by taking the judges' recommendations on the premises and inferring the court's decision on r via the rule $(p \land q) \leftrightarrow r$ that formalizes the contract law (*premise-based procedure*).[10] The problem is that the court's decision depends on the procedure adopted. In this specific example, under the conclusion-based procedure, the defendant will be declared not liable, whereas under the premise-based procedure, the defendant would be sentenced liable. As Kornhauser and Sager stated:

> We have no clear understanding of how a court should proceed in cases where the doctrinal paradox arises. Worse, we have no systematic account of the collective nature of appellate adjudication to turn to in the effort to generate such an understanding. [KS93, p. 12]

Legal theorists have discussed both methods and have taken different positions about them, either by arguing for the superiority of one of the approaches or by questioning both and recommending a third way (see Nash [Nas03] for an overview of the proposed solutions). In particular, Kornhauser and Sager argue against the use of a uniform voting protocol and favor instead a context-sensitive approach, where courts choose the method on a case-by-case basis, by voting on the method to be applied.

Discursive dilemma

Judgment aggregation has provided a systematic account of situations like the one arising in Figure 1.4. The first step was made by the political philosopher Pettit [Pet01], who recognized that

---

[10]The premise-based procedure has been reconsidered later as one of the possible escape routes from the many impossibility results that plague the discipline (see Section 4.3.1 later in the book).

	Valid contract	Breach	Legal doctrine	Defendant liable
	p	q	$(p \wedge q) \leftrightarrow r$	r
Judge 1	1	1	1	1
Judge 2	1	0	1	0
Judge 3	0	1	1	0
Majority	1	1	1	0

Figure 1.5: The discursive dilemma.

the paradox illustrates a more general problem than just an *impasse* in a court decision. Pettit introduced the term *discursive dilemma* to indicate any group decision in which the aggregation on the individual judgments depends on the chosen aggregation method, like the premise-based and the conclusion-based procedures.

Then, List and Pettit [LP04] reconstructed Kornhauser and Sager's example as shown in Figure 1.5. The difference with Figure 1.4 is that here the legal doctrine has been added to the set of issues on which the judges have to vote. Now the discursive dilemma is characterized by the fact that the group reaches an inconsistent decision, like $\{p, q, (p \wedge q) \leftrightarrow r, \neg r\}$. The court would accept the legal doctrine, give a positive judgment on both premises p and q but, at the same time, reach a negative opinion on the conclusion r. Clearly, such a position is untenable, as it would amount to release the defendant while saying, at the same time, that the two conditions for the defendant's liability applied.

What are the consequences of the reconstruction given in Figure 1.5? Mongin and Dietrich [MD10, Mon11] have investigated such reformulation and observed that:

> [T]he discursive dilemma shifts the stress away from the conflict of methods to *the logical contradiction within the total set of propositions that the group accepts*. [...] Trivial as this shift seems, it has far-reaching consequences, because all propositions are now being treated alike; indeed, the very distinction between premises and conclusions vanishes. This may be a questionable simplification to make in the legal context, but if one is concerned with developing a general theory, the move has clear analytical advantages. [Mon11, p. 2]

Indeed, instead of premises and conclusions, List and Pettit chose to address the problem in terms of *judgment sets*, i.e., the sets of propositions accepted by the individual voters. The theory of judgment aggregation becomes then a formal investigation on the conditions under which consistent individual judgment sets may collapse into an inconsistent collective judgment set.

Exactly like Arrow's theorem showed the full import of the Condorcet paradox, so showed the result of List and Pettit how far-reaching the doctrinal paradox and the discursive dilemma are. In the next section we will look at how the Condorcet paradox relates to these two paradoxes of the aggregation of judgments.

1.2.2   PREFERENCE AGGREGATION AND JUDGMENT AGGREGATION

Let us start by introducing some formal notation. Let X be a set of alternatives, and $\succ$ a binary predicate for a binary relation over X, where $x \succ y$ means "x is strictly preferable to y." The desired properties of preference relations viewed as strict linear orders are:

$$(P1) \quad \forall x, y((x \succ y) \rightarrow \neg(y \succ x)) \qquad \text{(asymmetry)}$$
$$(P2) \quad \forall x, y(x \neq y \rightarrow (x \succ y \vee y \succ x)) \quad \text{(completeness)}$$
$$(P3) \quad \forall x, y, z((x \succ y \wedge y \succ z) \rightarrow x \succ z) \quad \text{(transitivity)}$$

Example 1.1   Condorcet paradox as a doctrinal paradox   Suppose there are three possible alternatives x, y and z to choose from, and three voters V_1, V_2 and V_3 whose preferences are the same as in Figure 1.2. The three voters' preferences can then be represented by sets of preferential judgments as follows: $V_1 = \{x \succ y, y \succ z, x \succ z\}$, $V_2 = \{y \succ z, z \succ x, y \succ x\}$ and $V_3 = \{z \succ x, x \succ y, z \succ y\}$. According to Condorcet's method, a majority of the voters (V_1 and V_3) prefers x to y, a majority (V_1 and V_2) prefers y to z, and another majority (V_2 and V_3) prefers z to x. This leads us to the collective outcome $\{x \succ y, y \succ z, z \succ x\}$, which together with transitivity (P3) violates (P1) (Figure 1.6). Each voter's preference is transitive, but transitivity fails to be mirrored at the collective level. This is an instance of the Condorcet paradox casted in the form of a set of judgments over preferences on alternatives.[11]

What the Condorcet paradox and the discursive dilemma have in common is that when we combine individual choices into a collective one, we may lose some *rationality constraint* that was satisfied at the individual level, like transitivity (in the case of preference aggregation) or logical consistency (in the case of judgment aggregation). A natural question is then how the theory of judgment aggregation and the theory of preference aggregation relate to one another. We can address this question in two ways: we can consider what the possible interpretations are of aggregating judgments and preferences, and we can investigate the formal relations between the two theories.

On the first consideration, Kornhauser and Sager see the possibility of being right or wrong as the discriminating factor between judgments and preferences:

> When an individual expresses a preference, she is advancing a limited and sovereign claim. The claim is limited in the sense that it speaks only to her own values and advantage. The claim is sovereign in the sense that she is the final and authoritative arbiter

---

[11]We will come back later in Chapter 2 to another (logically simpler) formalization of the Condorcet paradox as a set of judgments about preferences (Example 2.15).

	$x \succ y$	$y \succ z$	$x \succ z$	$y \succ x$	$z \succ y$	$z \succ x$
V_1	1	1	1	0	0	0
V_2	0	1	0	1	0	1
V_3	1	0	0	0	1	1
Majority	1	1	0	0	0	1

Figure 1.6: The Condorcet paradox as a doctrinal paradox.

of her preferences. The limited and sovereign attributes of a preference combine to make it perfectly possible for two individuals to disagree strongly in their preferences without either of them being wrong. [...] In contrast, when an individual renders a judgment, she is advancing a claim that is neither limited nor sovereign. [...] Two persons may disagree in their judgments, but when they do, each acknowledges that (at least) one of them is wrong. [KS86, p. 85].[12]

Regarding the formal relations between judgment and preference aggregation, Dietrich and List [DL07a] (extending earlier work by List and Pettit [LP04]) capitalize on the representation of the Condorcet paradox given in Figure 1.6 and show that Arrow's theorem for strict and complete preferences can be derived from an impossibility result in judgment aggregation.[13]

Despite these natural connections, and the formal results they support, Kornhauser and Sager [Kor92] notice that the two paradoxes do not perfectly match. Indeed, as stated also by List and Pettit:

[W]hen transcribed into the framework of preferences instances of the discursive dilemma do not always constitute instances of the Condorcet paradox; and equally instances of the Condorcet paradox do not always constitute instances of the discursive dilemma. [LP04, pp. 216–217]

Given the analogy between the two paradoxes, List and Pettit's first question was whether an analogue of Arrow's theorem could be found for the judgment aggregation problem. Arrow showed that the Condorcet paradox hides a much deeper problem that does not affect only the majority rule. The same question could be posed in judgment aggregation: is the doctrinal paradox only the surface of a more troublesome problem arising when individuals cast judgments on a given set of propositions? As we shall see in more detail in Chapter 3, the answer to this question is positive and that can be seen as the starting point of the theory of judgment aggregation.

---

[12]Different procedures for judgment aggregation have been assessed with respect to their truth-tracking capabilities, see [BR06, HPS10].

[13]We will discuss this result in Chapter 3 (Section 3.4.1).

How likely are majority cycles?

Even from our brief survey, the reader may have guessed that large parts of the literature in social choice theory focused on the problem of majority cycles. We may wonder how likely such cycles are in reality. There are two main approaches to this question in the literature. One consists in analytically deriving the probability of a Condorcet paradox in an election, while the other looks at empirical evidence in actual elections. One assumption usually made in the first approach is the so-called *impartial culture*. According to the impartial culture, each preference ordering is equally possible. It should be noted that, even though it is a useful assumption for the analytic calculations, such an assumption has often been criticized as unrealistic. Niemi and Weisberg [NW68] showed that, under the impartial culture assumption and for a large number of voters, the probability of a majority cycle approaches 1 as the number of alternatives increases. However, they also found out that the probability of the paradox is quite insensitive to the number of voters but depends highly on the number of alternatives.

Yet, these results are in contrast with the findings of the approach that looks at the actual elections.[14] Mackie [Mac03], for example, claims that majority cycles never actually occurred in real elections. One way to explain such discrepancy is that we do not dispose of all the information needed to verify the occurrence of a majority cycle. For example, we typically do not dispose of the voter's preference order over all the possible candidates.

1.3   FURTHER TOPICS

The brief survey on social choice theory provided in this chapter has no pretense to be exhaustive. The aim was to give a background against which to frame the birth and development of judgment aggregation. For a broader but still concise introduction to social choice theory see [Lisce], and [Nur10, Pacds] for an introduction to voting theory. Moreover, the reader is referred to [RVW11] for a survey on preference reasoning from a perspective that brings together social choice theory and artificial intelligence. In particular, Chapter 4 of [RVW11] focuses on preference aggregation. There, several voting rules are defined, and manipulation and computational aspects are discussed. Manipulation in judgment aggregation is the object of a separate chapter in the present book (Chapter 5).

If the traditional domain of social choice theory has been economics and the political sciences, attention in aggregation problems is witnessing a steady growth within the fields of artificial intelligence and multi-agent systems. Aggregation problems often appear in the design and specification of distributed intelligent systems and the very same idea of voting has been applied to problems like recommender systems [PHG00] and rank aggregation for the Web [DKNS01]. In particular, computational social choice [CELM07, BCE13], of which judgment aggregation can be seen as a contributing field, is the discipline stemmed from the interactions between computer science and social choice theory, and which studies, among other topics, the computational

---

[14]See also [RGMT06] for an introduction to behavioral social choice.

complexity of the application and manipulation of aggregation rules [EGP12],[15] the design of aggregation rules based on knowledge representation techniques like merging [Pig06],[16] or the application of logic to reason, within a formal language, about aggregation problems [AvdHW11].

---

[15]We will touch upon this topic in Chapter 5 (Section 5.3.3).
[16]We will discuss this topic in detail in Chapter 4 (Section 4.3.3) and Chapter 6.

CHAPTER 2

Basic Concepts

This chapter is devoted to an introduction of the basic framework of judgment aggregation based on propositional logic. Our presentation is based on the framework first proposed in [LP02] and later developed by Dietrich and List in a long series of works (e.g., [DL07a, DL07c] to name just a few).

Chapter outline: We start in Section 2.1 by introducing the notions of agenda, judgment set, judgment profile, and aggregation function. In the same section we will also define a number of concrete aggregation functions. Section 2.2 proceeds by defining some properties of agendas, which have to do with how 'tightly' the formulae in the agenda are logically related to one another. We will see later that the more interconnected an agenda is, the more difficult the aggregation problem becomes. In Section 2.3 we look into a set of natural properties that one might wish to impose on the aggregation function to guarantee its 'good' behavior. In the concluding section we refer the reader to alternative formal frameworks—not necessarily based on logic—that have been developed in the literature to cast the theory of judgment aggregation.

2.1   PRELIMINARIES

2.1.1   AGENDAS IN PROPOSITIONAL LOGIC

In this book we will only be concerned with the aggregation of judgments that are expressed in propositional logic, which has been the framework of choice for most of the literature.[1] So we start by briefly recapitulating—for the readers unfamiliar with propositional logic—some basic notions from its syntax and semantics. For a comprehensive exposition the reader is referred to [vD80, Ch. 1].

Propositional logic
The language of propositional logic, which we denote by $\mathcal{L}$, consists of all the formulae that can be defined inductively from a countable set $At = \{p, q, \ldots\}$ of atomic propositions (also called atoms) using the logical connectives ¬ (negation), ∧ (conjunction), ∨ (disjunction), → (implication), ↔ (equivalence). The inductive definition goes as follows: [*Base*] all elements of At are formulae in $\mathcal{L}$; [*Step*] if φ and ψ belong to $\mathcal{L}$, then also $\neg\varphi$ ("not φ"), $\varphi \wedge \psi$ ("φ and ψ"), $\varphi \vee \psi$ ("φ or ψ"), $\varphi \rightarrow \psi$ ("if φ then ψ"), and $\varphi \leftrightarrow \psi$ ("φ if and only if ψ") belong to $\mathcal{L}$, and nothing else belongs

---

[1]The only exception will be Chapter 7. The case of richer logics will be touched upon in the concluding section of this chapter.

to $\mathcal{L}$.[2] We say that a formula is *positive* if its outermost connective is not a negation (e.g., $p \rightarrow q$, $\neg p \vee q$).[3]

The meaning of a formula $\varphi \in At$ is its *truth value* as specified by a *valuation function* $\mathcal{V} : \mathcal{L} \longrightarrow \{0, 1\}$ where 0 stands for "false" and 1 for "true." Each valuation $\mathcal{V}$ is an extension of some valuation $\mathcal{V} : At \longrightarrow \{0, 1\}$ of truth values to atoms, which obeys the following constraints: $\mathcal{V}(\neg\varphi) = 1 - \mathcal{V}(\varphi)$; $\mathcal{V}(\varphi \wedge \psi) = 1$ iff $\mathcal{V}(\varphi) = \mathcal{V}(\psi) = 1$; $\mathcal{V}(\varphi \vee \psi) = 0$ iff $\mathcal{V}(\varphi) = \mathcal{V}(\psi) = 0$; $\mathcal{V}(\varphi \rightarrow \psi) = 1$ iff $\mathcal{V}(\varphi) = 0$ or $\mathcal{V}(\psi) = 1$; $\mathcal{V}(\varphi \leftrightarrow \psi) = 1$ iff $\mathcal{V}(\varphi) = \mathcal{V}(\psi)$. These constraints define the *semantics* of the logical connectives introduced above. When $\mathcal{V}(\varphi) = 1$ (respectively, $\mathcal{V}(\varphi) = 0$) we will often write $\mathcal{V} \models \varphi$ (respectively, $\mathcal{V} \not\models \varphi$). If Φ is a set of formulae, we write $\mathcal{V} \models \Phi$ to express that for all $\varphi \in \Phi$, $\mathcal{V} \models \varphi$, i.e., all formulae in φ are made true by $\mathcal{V}$.

We conclude with some auxiliary terminology concerning special classes of propositional formulae. A formula φ is a *tautology* if, for any valuation $\mathcal{V}$, $\mathcal{V} \models \varphi$; it is a *contradiction* if, for any valuation $\mathcal{V}$, $\mathcal{V} \not\models \varphi$; it is *contingent* if it is neither a tautology nor a contradiction. A set of formulae Φ is *consistent* if it has a model, that is, if there exists a valuation $\mathcal{V}$, such that $\mathcal{V} \models \varphi$ for each $\varphi \in \Phi$; a formula φ is a *logical consequence* of a set of formulae Φ (in symbols, $\Phi \models \varphi$) if for every valuation $\mathcal{V}$ such that $\mathcal{V} \models \Phi$, it is the case that $\mathcal{V} \models \varphi$.

Agendas

With the machinery of propositional logic in place, we can frame the problem of the aggregation of judgments simply as a set of individuals or agents that are called to decide upon a given set of issues:

Definition 2.1 Judgment aggregation problem.   Let $\mathcal{L}$ be a propositional language on a given set of atoms At. A *judgment aggregation problem* for $\mathcal{L}$ is a tuple $\mathcal{J} = \langle N, A \rangle$ where:

- N is a finite non-empty set;

- $A \subseteq \mathcal{L}$ such that $A = \{\varphi \mid \varphi \in I\} \cup \{\neg\varphi \mid \varphi \in I\}$ for some finite $I \subseteq \mathcal{L}$ which contains only positive contingent formulae.

Set N is the set of *individuals* (or *agents* or *voters*). A is called the *agenda* and I is called the set of *issues* or the *pre-agenda* of A. An agenda based on a set of issues I will often be denoted $\pm I$. Given an agenda A, we denote its pre-agenda by $[A]$.[4]

Intuitively, one can view a judgment aggregation problem as what specifies the space of possible situations in which the individuals in N have to reach some collective decision about the issues in I. An agenda $A = \pm I$ represents then all possible attitudes that can be assumed toward

---

[2]More compactly, $\mathcal{L}$ is defined by the grammar: $\varphi := p \in At \mid \neg\varphi \mid \varphi \wedge \varphi \mid \varphi \vee \varphi \mid \varphi \rightarrow \varphi \mid \varphi \leftrightarrow \psi$.

[3]We will assume the standard notational conventions for the relative strength of the binding of connectives ($\wedge$ and $\vee$ bind more strongly than $\rightarrow$ and $\leftrightarrow$, and $\neg$ binds more strongly than all other connectives) and for the use of brackets.

[4]Clearly, $[\pm I] = I$.

the issues in I. In the framework we are going to work with, such attitudes are of only two types: acceptance and rejection. The agenda is therefore a set of formulae which is closed under negation, i.e., $\forall \varphi: \varphi \in A$ iff $\neg \varphi \in A$, and where double negations are eliminated. To make an example, the doctrinal paradox agenda $\{p, q, p \wedge q\} \cup \{\neg p, \neg q, \neg(p \wedge q)\} = \pm \{p, q, p \wedge q\}$ expresses all the acceptance/rejection attitudes that one individual can assume over the set of issues $\{p, q, p \wedge q\}$.

2.1.2   JUDGMENT SETS AND PROFILES

Given a judgment aggregation problem, individuals are asked to express their opinions on the formulae of the agenda by accepting some and rejecting others. These opinions are called *judgment sets* and are defined as follows:

Definition 2.2   Judgment sets and profiles.   Let $\mathcal{J} = \langle N, A \rangle$ be a judgment aggregation problem. A *judgment set* for $\mathcal{J}$ is a set of formulae $J \subseteq A$ such that:

- J is consistent;

- J is complete, i.e., $\forall \varphi \in A$, either $\varphi \in J$ or $\neg \varphi \in J$.

Instead of $\varphi \in J$ we will often use the notation $J \models \varphi$ to indicate that φ belongs to judgment set J.[5] The set of all judgment sets is denoted $\mathbf{J} \subseteq \wp(A)$, where $\wp(.)$ denotes the power-set function. A *judgment profile* $P = \langle J_i \rangle_{i \in N} \in \mathbf{J}^{|N|}$ is an $|N|$-tuple of judgment sets. With P_i we denote the ith entry of P, i.e., the judgment set of agent i in P. For $\varphi \in A$, we use P_φ to denote the set of individuals accepting φ in P: $\{i \in N \mid P_i \models \varphi\}$. Finally, we denote with $\mathbf{P}$ the set of all judgment profiles. Abusing notation, we will sometimes indicate that a judgment set J_i belongs to a profile P by writing $J_i \in P$.

So individuals express their opinions through sets of formulae of the agenda: the formulae contained in the set are the ones that are *accepted* by the individual, the ones belonging to the complement of the set are the ones that are *rejected* by the individual. The consistency and completeness criteria formalize a notion of 'rationality' for the views that might be held by individuals. Such views cannot be internally contradictory (consistency) and cannot abstain from accepting or rejecting any of the issues posed by the agenda (completeness).[6]

Remark 2.3   Deductive closure   A set of formulae Φ is *deductively* closed (w.r.t. agenda A) if any $\varphi \in A$ that follows logically from Φ is also contained in it: if $\Phi \models \varphi$, then $\varphi \in \Phi$. Since judgment sets are sets of formulae that are consistent and complete, they are also deductively closed. However, a set of formulae that is consistent and deductively closed is not necessarily complete. When working with judgment sets the two notations $\varphi \in J$ (membership) and $J \models \varphi$

---

[5]Clearly, if $J \subseteq A$ is a judgment set then: $\varphi \in J$ iff $J \models \varphi$. This is not true in general for any subset of formulae $\Phi \subseteq A$ of the agenda. See Remark 2.3 below.

[6]This is true for the standard theory of judgment aggregation. However, as we will see in Chapter 4, Section 4.2, some work considered to relax the completeness so to allow individuals to abstain on some of the agenda's issues.

(consequence) can be seen as notational variants. However, when working with sets of formulae that are not judgment sets by the letter of Definition 2.2—in our context these will typically be sets of formulae accepted by a group of individuals—we will keep the two notations distinct.

2.1.3   AGGREGATION FUNCTIONS

The judgment aggregation problem consists in the aggregation of the individuals' judgment sets into one collective judgment set. The aggregation of individual judgments is viewed as a function:

Definition 2.4   Aggregation function.   Let $\mathcal{J} = \langle N, A \rangle$ be a judgment aggregation problem. An *aggregation function* for $\mathcal{J}$ is a function $f : \mathbf{P} \longrightarrow \wp(A)$. The output set $f(P)$, where $P = \langle J_i \rangle_{i \in N}$, is sometimes denoted J. Set J is then called a *collective set*. A collective set J which is a judgment set is called a *collective judgment set*.

So, an aggregation function takes as input a profile of consistent and complete subsets of the agenda (i.e., judgment sets) and outputs a subset of the agenda. Such subset is neither necessarily consistent nor necessarily complete. In other words, the collective set is not necessarily a judgment set. In view of our discussion of the doctrinal paradox and the discursive dilemma this should not come as a surprise: the output of an aggregation function might not be 'rational' in the sense in which individual judgment sets are.

Remark 2.5   Universal domain and resoluteness   We conclude our comment of Definition 2.4 by noticing that it builds two key properties into the notion of aggregation function. First, it assumes that the domain of the aggregation consists of all possible judgment profiles or, intuitively, that all profiles of individual opinions are admissible as input for the aggregation. This property is commonly referred to as *universal domain*. Second, it assumes the aggregation to be *resolute*, that is, to yield for each profile only one set of formulae. In this book we will work almost exclusively with functions that satisfy universal domain and resoluteness. Aggregation functions that do not satisfy universal domain will be presented later in Chapter 4. Irresolute functions yielding for each profile a non-empty set of sets of formulae will be studied later in Chapter 4 and especially in Chapter 6.

2.1.4   EXAMPLES: AGGREGATION RULES

We now give several examples of aggregation functions as rules for defining the collective set based on a judgment profile. We typically refer to concrete aggregation functions as *aggregation rules*. The ones that follow in this section will be discussed at several places throughout the book and are the ones most commonly considered in the literature.

Threshold-based rules

The rules below determine the collective outcome by checking, for each proposition in the agenda (they are therefore commonly referred to as *propositionwise* rules), whether the number of individuals accepting that formula exceeds a given threshold. If that is the case, the formula is collectively accepted. Let $P \in \mathbf{P}$, we define the following rules.

Majority rule:

$$f_{maj}(P) = \left\{ \varphi \in A \mid |P_\varphi| \geq \left\lceil \frac{|N| + 1}{2} \right\rceil \right\} \qquad (2.1)$$

where, for $x \in \mathbb{Q}$, $\lceil x \rceil$ is the smallest integer greater or equal to x. I.e., φ is collectively accepted iff there is a majority of individuals accepting it. We will refer to this rule as the *propositionwise majority rule* or simply as the *majority rule*.

Unanimity rule:

$$f_u(P) = \left\{ \varphi \in A \mid |P_\varphi| \geq |N| \right\} \qquad (2.2)$$

I.e., φ is collectively accepted iff all individuals accept it. We will refer to this rule as the *propositionwise unanimity rule* or simply as the *unanimity rule*.

Quota rule:

$$f_t(P) = \left\{ \varphi \in A \mid |P_\varphi| \geq t_\varphi \right\} \qquad (2.3)$$

where $t = \langle t_\varphi \rangle_{\varphi \in A}$ is a tuple of integer thresholds or quotas t_φ, one for each formula in the agenda. I.e., φ is collectively accepted iff there are at least t_φ individuals that accept it.

Formula 2.3 defines the class of all propositionwise threshold-based rules. Clearly, the propositionwise majority rule is a particular quota rule whose threshold has been fixed at $\left\lceil \frac{|N|+1}{2} \right\rceil$ for all formulae in the agenda.[7] Similarly, the unanimity rule is a quota rule with threshold $|N|$ for all formulae. Quota rules that assign the same threshold to all formulae called *uniform*.

It must be noted that the selection of the thresholds has an impact on the 'rationality' of the collective set. For instance, it is not difficult to see that the unanimity rule might return incomplete collective sets, and that a uniform quota rule imposing a common threshold lower than $\left\lceil \frac{N}{2} \right\rceil$ might return collective sets containing both a formula and its negation. In general, one can identify precise constraints on the thresholds, which can enforce a well-behaved output of the aggregation. For instance, for each pair φ and $\neg\varphi$, the inequalities

$$t_\varphi + t_{\neg\varphi} \quad \leq \quad |N| + 1 \qquad (2.4)$$
$$t_\varphi + t_{\neg\varphi} \quad \geq \quad |N| \qquad (2.5)$$

---

[7]In voting theory, the majority rule is indeed called *simple majority rule* to distinguish it from supermajority (or qualified) majority requiring a support greater than 50% of the individuals.

are necessary and sufficient conditions for the collective set to be complete (i.e., to contain at least one of φ or $\neg\varphi$, Formula 2.4) and, respectively, to be such that it never contains both a formula and its negation (i.e., to contain at most one of φ and $\neg\varphi$, Formula 2.5). This latter property is usually referred to as *weak consistency*.

The class of all quota rules has been studied extensively in [DL07b]. We will come back to the majority rule in much more detail later in Chapter 3, and to quota rules as possible escape routes to some of the impossibility results of judgment aggregation in Chapters 4 and 5.[8]

Premise- and conclusion-based rules

We have already encountered the premise- and conclusion-based rules in Section 1.2. Here we give a more precise formulation of them.

Premise-based rule:

$$f_{pb}(P) = f_{maj}(P^{Prem}) \cup \{\varphi \in Conc \mid f_{maj}(P^{Prem}) \models \varphi\} \tag{2.6}$$

where: $Prem \subseteq A$ consists of the subagenda containing the issues that are considered *premises* in the aggregation, and their negations; $Conc \subseteq A$ consists of the subagenda containing the issues that are considered *conclusions* in the aggregation, and their negations; $Prem$ and $Conc$ are a partition of A; and P^{Prem} (respectively, P^{Conc}) denotes the profile obtained from the restrictions of the judgment sets to the formulae in $Prem$ (respectively, $Conc$). I.e., φ is collectively accepted iff it is a premise and it has been voted by the majority of the individuals or it is a conclusion entailed by the premises accepted by the majority.

Conclusion-based rule:

$$f_{cb}(P) = f_{maj}(P^{Conc}) \tag{2.7}$$

where P^{Conc} is as for the premise-based rule. I.e., φ is collectively accepted iff it is a conclusion and it has been voted by the majority of the individuals.

Intuitively, premise- and conclusion-based rules apply propositionwise aggregation, via the majority rule, only to specific subsets of the agenda, viz., its premises or its conclusions. They have played a pivotal role in the development of the theory of judgment aggregation, and much literature has been dedicated to their analysis (see, for instance, [NP06, DM10]). We will come back to them at several places in the remaining of the book.

An example

It is now time to illustrate the workings of all the above rules side by side. We do that with yet another variant of the doctrinal paradox:

Example 2.6   Let $A = \pm\{p, p \to q, q\}$, and attach the following intuitive reading to the three issues [DL07a]:

---

[8]See in particular Sections 4.2.2 and 5.3.

	p	$p \to q$	q
J_1	1	1	1
J_2	1	0	0
J_3	0	1	0
f_{maj}	1	1	0
f_u			
$f_{t'}$	1	1	0
$f_{t''}$	1	0	0
f_{pb}	1	1	1
f_{cb}			0

Figure 2.1: An illustration of several aggregation rules from Example 2.6.

p: Current CO_2 emissions lead to global warming.
$p \to q$: If current CO_2 emissions lead to global warming, then we should reduce CO_2 emissions.
q: We should reduce CO_2 emissions.

The profile consisting of the three judgment sets $J_1 = \{p, p \to q, q\}$, $J_2 = \{p, \neg(p \to q), \neg q\}$ and $J_3 = \{\neg p, p \to q, \neg q\}$, once aggregated via propositionwise majority (f_{maj}), gives rise to an inconsistent collective judgment set $J = \{p, p \to q, \neg q\}$. Propositionwise unanimity (f_u) does not accept any of the items of the agenda. If we assume that $Prem = \{p, p \to q\}$ and $Conc = \{q\}$, the premise-based rule (f_{pb}) generates a collective judgment accepting all items, and the conclusion-based rule (f_{cb}) rejects the conclusion q, and does not accept any other item of the agenda.

We give two examples of quota rules. The first is a quota rule that requires majority over the premises and their negations, but requires a unanimous vote to collectively accept the positive conclusion and one individual to reject it. That is: $t'_p = t'_{p \to q} = t'_{\neg p} = t'_{\neg(p \to q)} = \left\lceil \frac{|N|+1}{2} \right\rceil$, $t'_q = |N|$, and $t'_{\neg q} = 1$.[9] This quota rule accepts both premises but rejects the conclusion. The second one requires majority on all atomic issues and unanimity on the implicative issues.[10] That is: $t''_p = t''_{\neg p} = t''_q = t''_{\neg q} = \left\lceil \frac{|N|+1}{2} \right\rceil$, $t''_{p \to q} = |N|$ and $t''_{\neg(p \to q)} = 1$.[11] This rule then accepts one premise but rejects the implicative premise and the conclusion. Figure 2.1 recapitulates the outputs just discussed.

It is no accident that all aggregation rules in the above example either fail to yield a judgment set (all except f_{pb} and $f_{t''}$, whose output is consistent and complete) or output sets that are

---

[9]Notice that these thresholds satisfy the constraints in Formulae 2.4 and 2.5.
[10]Thresholds of this type, on agendas containing only the implication connective are extensively studied in [Die10].
[11]Again, note that these thresholds satisfy the constraints in Formulae 2.4 and 2.5.

inconsistent with one another (f_{pb} accepts q while f_{cb} and $f_{t''}$ reject it). The reasons for such failure are deep and we will probe them in Chapter 3. The remainder of the present chapter sets the stage for those investigations.

2.2 AGENDA CONDITIONS

We introduce here three conditions on agendas, which capture the sort of logical interdependence possibly arising between their elements.

2.2.1 HOW INTERCONNECTED IS AN AGENDA?

We define and illustrate the agenda conditions known as non-simplicity, even-negatability and path-connectedness. We also introduce the auxiliary notion of conditional entailment.

Non-simplicity

The first agenda condition is almost self-explanatory, and is usually referred to as *non-simplicity* [NP07].

Definition 2.7 Non-simple agendas.   An agenda A is *non-simple* (NS) iff it contains at least one set X s.t.:

- $3 \le |X|$;

- X is *minimally inconsistent*, i.e.:

  - X is inconsistent;
  - $\forall Y$ s.t. $Y \subset X$: Y is consistent.

An agenda is called *simple* if it is not non-simple.

It is easy to see that agenda $\pm\{p, q, p \wedge q\}$ is non-simple as the set $\{p, q, \neg(p \wedge q)\}$ is clearly minimally inconsistent. Notice that if X is minimally inconsistent then, for some $\varphi \in X$, it is not only the case that $X - \{\varphi\}$ is consistent, but also that $X - \{\varphi\} \models \neg\varphi$. Non-simplicity is the minimal level of complexity for an agenda to run into problems when attempting aggregation.

On the other hand, we will see that if an agenda is simple then aggregations of a non-degenerate kind are possible. In fact, the propositionwise majority rule can be proven to be the unique aggregation function that satisfies some highly desirable properties.[12] By Definition 2.7, simple agendas are agendas where minimally inconsistent sets have cardinality of at most two.[13] Examples are agendas whose issues consist of logically unrelated formulae (e.g., $\{p, q, r\}$),[14] or agendas whose issues can be ordered by logical strength like $\pm\{p, p \wedge q, p \wedge q \wedge r\}$.

---

[12]We will come back to this result (Theorem 3.2) later in Section 3.4.

[13]Simplicity was introduced by [NP07] as the *median property*, i.e., the property according to which every minimally inconsistent subset of the agenda has size ≤ 2.

[14]These agendas are also known as *bipolar agendas* [DL10b].

Conditional entailment

From non-simplicity we move now to the related notion of *conditional entailment* [DL13a]. This will be needed later to define the condition of path-connectedness.

Definition 2.8   Conditional entailment.   Let $\varphi, \psi \in A$. We say that φ *conditionally entails* ψ (notation: $\varphi \models_c \psi$) if for some (possibly empty) $X \subseteq A$, which is consistent with φ and with $\neg\psi$, $\{\varphi\} \cup X \models \psi$.

Conditional entailment expresses that the acceptance of ψ follows from the acceptance of φ either directly—by logical consequence—or indirectly once a set of formulae X is also accepted, which is compatible with both the acceptance of φ and the rejection of ψ. Intuitively, the fact that φ conditionally implies ψ captures a specific dependency within the structure of the agenda whereby if, on the one hand, it is possible to accept both φ and the formulae in X or both X and $\neg\psi$, on the other hand, accepting φ and X would compel one to also accept ψ.

A few observations are in order. If $\varphi \models \psi$ (i.e., ψ is a logical consequence of φ) then $\varphi \models_c \psi$ since ψ follows from $\{\varphi\} \cup \emptyset$. If an agenda contains $\varphi \neq \psi$ such that $\varphi \models_c \psi$, then the agenda must have been generated by a set of issues containing at least two formulae. We conclude with the following observation relating conditional entailment to the property of non-simplicity:

Fact 2.9   Conditional entailment and NS   Let A be an agenda and $\varphi, \psi \in A$. If (i) $\varphi \models_c \psi$ and (ii) $\varphi \not\models \psi$, then A satisfies NS.

Proof. By *(i)*, *(ii)* and Definition 2.8 it follows that there exists $X \neq \emptyset$ such that $\{\varphi\} \cup X \models \psi$ and hence such that $X \cup \{\varphi, \neg\psi\}$ is inconsistent. By the compactness[15] of propositional logic there exists a smallest non-empty X' such that $X' \cup \{\varphi, \neg\psi\}$ is inconsistent. Set $X' \cup \{\varphi, \neg\psi\}$ is therefore minimally inconsistent and has cardinality bigger or equal to 3. Hence A satisfies NS.
□

Even negations

The second agenda condition is known as even negatability or even number negations property, and is slightly more involved:

Definition 2.10   Evenly negatable agendas.   An agenda A satisfies the *even negations* condition (EN) iff:

- A contains a minimally inconsistent set $X \subseteq A$ and a set $Y = \{\varphi, \psi\} \subseteq X$, such that $X - Y \cup \{\neg\varphi, \neg\psi\}$ is consistent.[16]

---

[15]The compactness of propositional logic guarantees that, if $X \models \varphi$ (with X possible infinite), then there exists a smallest (finite) set $X' \subseteq X$ such that $X' \models \varphi$.

[16]Notice that, technically, $X - Y \cup \{\neg\varphi, \neg\psi\}$ is not necessarily a subset of the agenda because it might contain double negations. Clearly, the removal of double negations yields an equivalent set that is a subset of the agenda.

In other words, the agenda satisfies the even number negations condition whenever it contains at least one minimally inconsistent set X that can be made consistent by negating two of its elements.[17] Notice that no cardinality restriction is imposed on X itself.

Again, it is easy to see that the agenda $\pm\{p, q, p \wedge q\}$ satisfies this property, as well as $A = \pm\{p, q, p \to q\}$. Other agendas do not:

Example 2.11   Non evenly negatable agendas   Consider $A = \pm\{p, q, p \leftrightarrow q\}$. We have the following minimally inconsistent sets: $\{p, \neg q, p \leftrightarrow q\}$, $\{\neg p, q, p \leftrightarrow q\}$, $\{p, q, \neg(p \leftrightarrow q)\}$, $\{\neg p, \neg q, \neg(p \leftrightarrow q)\}$. None of these sets can be made consistent by negating any pair of formulae in the set.

Example 2.12   Evenly negatable agendas (Arrow's agenda)   Consider the following agenda, which we have already encountered in Chapter 1: $\pm\{a \prec b, b \prec c, c \prec a\}$ where $a \prec b$, $b \prec c$ and $c \prec a$ are taken to be atomic propositions. Assume now the agenda is constrained by the set of propositional formulae so defined for $x \neq y \neq z \in \{a, b, c\}$:

$$\begin{array}{ll} \text{(asymmetry)} & x \succ y \to \neg(y \succ x) \\ \text{(completeness)} & (x \succ y \vee y \succ x) \wedge \neg(x \succ y \wedge y \succ x) \\ \text{(transitivity)} & (x \succ y \wedge y \succ z) \to x \succ z \end{array}$$

The constraints reproduce, in propositional form, the first-order constraints P1-P3 of linear orders we have already encountered in Section 1.2.2. Under these constraints, the agenda satisfies EN since set $\{a \succ b, b \succ c, c \succ a\}$ is minimally inconsistent and can be made consistent by swapping the first and third elements.

Path-connectedness

The third agenda condition is known as *path-connectedness* [DL13a] and was first introduced in [NP02] under the name of *total-blockedness*.

Definition 2.13   Path-connected agendas.   An agenda A is *path-connected* (PC) iff for all $\varphi, \psi \in A$ there exists a sequence $\varphi_1, \ldots, \varphi_n$ of elements of A s.t.: $\varphi = \varphi_1$, $\psi = \varphi_n$ and $\varphi_i \models_c \varphi_{i+1}$ for $1 \leq i < n$.

So, we call an agenda path-connected whenever any two of its formulae are logically connected in either a direct way, or in an indirect way by fixing the truth value of some other formula in the agenda. Intuitively, path-connectedness expresses a 'tightness' condition over agendas

---

[17]The condition is more often stated in the (apparently weaker) version requiring Y to be of even size instead of being of size 2. The two formulations are however equivalent as shown in [DL13a]. In [DL07a] the condition is referred to as *minimal connectedness*, in [Lis12] as *even number negatability*. It is known to be equivalent to the non-affineness condition introduced in [DH10a].

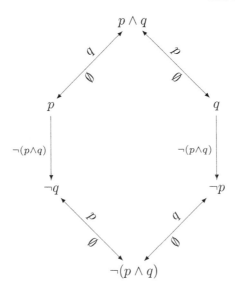

Figure 2.2: A directed graph representing the conditional entailment relations in agenda $\pm\{p, q, p \wedge q\}$ (reflexive arrows are omitted). The three lower elements cannot be connected to the three upper elements. An arrow from φ to ψ denotes $\varphi \models_c \psi$. If an arrow from φ to ψ is labeled by χ it means that $\varphi \cup \chi \models \psi$ and $\{\varphi, \chi\}$ and $\{\chi, \neg\psi\}$ are consistent. The relevant label for an arrow from φ to ψ is the one closer to ψ.

whereby the acceptance/rejection of an issue can, under different conditions, demand the acceptance/rejection of any other issue. In yet other words, there exists a $\models_c$-path connecting any two elements in the agenda.[18] Path-connectedness is a demanding agenda condition. Here are two examples:

Example 2.14 Path-*dis*connected agendas The agenda of the doctrinal paradox $\pm\{p, q, p \wedge q\}$ is not path-connected. This can be appreciated by noticing that no negative proposition conditionally entails a positive proposition in the agenda. Figure 2.2 displays a directed graph depicting the conditional entailment relations for this agenda: a directed arrow from φ to ψ indicates that φ conditionally implies ψ. Arrows are labeled with the formulae that establish the conditional entailment in the relevant direction (e.g., $p \models_c p \wedge q$ with q being the extra assumption determining the entailment). Similar considerations can be made for agendas $\pm\{p, q, p \to q\}$ and $\pm\{p, q, p \vee q\}$.[19]

---

[18]It might be instructive to also notice that PC is equivalent to the requirement that the transitive closure of the conditional entailment relation of Definition 2.8 covers the Cartesian square of the agenda, i.e., $(\models_c)^+ = A \times A$, where $(\models_c)^+$ denotes the transitive closure of $\models_c$.

[19]Recall that in propositional logic $p \to q$ is equivalent to $\neg(p \wedge \neg q)$ and $p \vee q$ is equivalent to $\neg(\neg p \wedge \neg q)$.

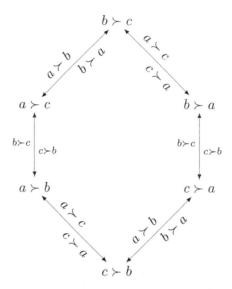

Figure 2.3: A directed graph representing (some of) the conditional entailment relations in agenda $\pm\{a \prec b, b \prec c, c \prec a\}$. Again, the relevant labels for reading the conditional entailments off the arrows are the ones closer to the head of the arrow, and reflexive arrows are omitted. Each formula is reachable through a path from any other formula.

Example 2.15   Path-connected agendas Consider again Arrow's agenda $\pm\{a \prec b, b \prec c, c \prec a\}$ from Example 2.12. This agenda satisfies PC, as its conditional entailment graph in Figure 2.3 shows. Another agenda satisfying PC, which we have encountered in the previous section is the discursive dilemma agenda: $\pm\{p, q, r, r \leftrightarrow (p \wedge q)\}$.

2.2.2   COMPARING AGENDA CONDITIONS

The above agenda conditions seem to impose rather different constraints on the strength of interconnections between the elements of the agenda. It is therefore worth looking at how they logically relate to one another:

Fact 2.16   Relative strength of agenda conditions

  i) EN, NS and PS have a non-empty intersection;

 ii) EN and NS are logically independent;

iii) PC and EN are logically independent;

iv) PC implies NS.

Proof. *i)* Arrow's agenda $\pm\{a \prec b, b \prec c, c \prec a\}$ (recall Examples 2.12 and 2.15) is non-simple, evenly negatable and path-connected. Also the agenda of the discursive dilemma $\pm\{p, q, r, r \leftrightarrow (p \wedge q)\}$ satisfies all three conditions. *ii)* There exist agendas which satisfy NS but not EN (e.g., $\pm\{p, q, p \leftrightarrow q\}$) and agendas which satisfy EN but not NS (e.g., $\pm\{p, r, p \wedge q\}$). *iii)* There exist agendas which satisfy PC but not EN (e.g., $\pm\{p, q, p \leftrightarrow q\}$ again), and agendas which satisfy EN but not PC (e.g., $\pm\{p, q, p \rightarrow q\}$). *iv)* Assume PC and take φ, ψ such that $\varphi \models_c \psi$. We have two cases: either (a) $\varphi \not\models \psi$ or (b) $\varphi \models \psi$. If (a), then by Fact 2.9, the agenda satisfies NS. If (b), then $\neg\varphi \not\models \psi$, since φ and ψ are contingent (Definition 2.1). By PC, there exists a chain of conditional entailments from $\neg\varphi = \psi_1$ to $\varphi = \psi_n$. We have in particular that $\psi_i \models_c \psi_{i+1}$ and $\psi_i \not\models \psi_{i+1}$, for $1 \leq i < n$. By Fact 2.9, we thus obtain that the agenda satisfies NS. $\square$

This concludes our presentation of the structural conditions on agendas most commonly considered in the judgment aggregation literature.[20] We now move to a discussion of the conditions that can be imposed on the aggregation function.

2.3 AGGREGATION CONDITIONS

How would we like an aggregation function to behave? For instance, do we wish the aggregation to be such that if all individuals accept one issue, then also their aggregate does? Do we wish all individuals to have the same weight in the aggregation, and that the aggregation treats the acceptance and the rejection of each issue in an unbiased way? This section introduces formal definitions of these—and many other—properties of aggregation functions that one might like to impose on aggregation functions in order to constrain their behavior.

Before starting, let us first fix some auxiliary terminology and notation, which will help us to streamline our exposition in the coming sections. Let $\Phi, \Phi' \subseteq A$ and $P, P' \in \mathbf{P}$:

- Φ *agrees with* Φ' on formula φ (notation: $\Phi =_\varphi \Phi'$) whenever it is the case that $\varphi \in \Phi$ if and only if $\varphi \in \Phi'$;

- P is an i-*variant* of P' (notation: $P =_{-i} P'$) whenever P and P' differ possibly only in their i^{th} entries, that is, $\forall j \neq i : P_j = P'_j$.

---

[20]While the agenda conditions we have presented are among the ones that have most commonly been used in the literature, many others have been studied. A comprehensive overview can be found in [DL13a].

By $\Phi \neq_\varphi \Phi'$ we indicate that Φ does not agree with Φ' on φ,[21] and by $P \neq_{-i} P'$ that P and P' are not i variants of one another.

2.3.1   HOW SHOULD AN AGGREGATION FUNCTION BEHAVE?

We are now ready to list a number of aggregation conditions that are commonly contemplated in the literature on judgment aggregation. Aggregation conditions are properties of aggregation functions. Aggregation functions represent different procedures for aggregating judgments. So the question arises of how to compare them and, crucially, of how to justify the application of one over another. The more 'good' properties are satisfied by a function, one might say, the better. The exact formulation of aggregation conditions allow us to compare different aggregation functions and study their behavior formally.

We identify two sets of conditions: *output conditions* expressing properties of the output of an aggregation function; and *mapping conditions* expressing properties of how the input is mapped to the output of the aggregation.[22] We will comment extensively on the definitions of these conditions in the following section.

Definition 2.17   Output conditions.   Let $\mathcal{J} = \langle N, A \rangle$ be a judgment aggregation problem and $X \subseteq A$. An aggregation function f for $\mathcal{J}$ is:

Consistent iff $\forall P \in \mathbf{P}$, $f(P)$ is a consistent set of formulae.
>   I.e., the collective set expresses a logically consistent view.

Complete iff $\forall \varphi \in A, \forall P \in \mathbf{P}, \varphi \in f(P)$ OR $\neg\varphi \in f(P)$.
>   I.e., the collective set is not undecided about any issue.

Closed iff $\forall \varphi \in A, \forall P \in \mathbf{P}$, IF $f(P) \models \varphi$ THEN $\varphi \in f(P)$.
>   I.e., the collective set accepts all logical consequences (contained in the agenda) of the formulae it accepts.

Collectively rational or simply **rational** (**RAT**) iff $f(P)$ is consistent and complete.
>   I.e., the collective set is a judgment set.

So an aggregation function is called rational if and only if it always outputs consistent and complete sets of formulae, that is, if and only if it always outputs judgment sets. To say it otherwise, the aggregation is rational whenever its type is $f : \mathbf{P} \longrightarrow \mathbf{J}$.

Definition 2.18   Mapping conditions.   Let $\mathcal{J} = \langle N, A \rangle$ be a judgment aggregation problem. An aggregation function f for $\mathcal{J}$ is:

---

[21]Notice that if Φ is a judgment set (that is, it is consistent and complete) accepting φ and Φ' is the collective set resulting from an aggregation (hence not necessarily consistent and complete), then $\Phi \neq_\varphi \Phi'$ may indicate either that $\neg\varphi \in \Phi'$, or that neither φ nor $\neg\varphi$ belong to Φ'.

[22]The latter are sometimes also referred to as responsiveness conditions [Lis12].

Anonymous (AN) iff $\forall P, P' \in \mathbf{P}$ s.t. P' is a permutation of P:[23] $f(P) = f(P')$.

I.e., all individuals have the same weight in the aggregation.

Unanimous (U) iff $\forall \varphi \in A, \forall P \in \mathbf{P}$: IF $[\forall i \in N : P_i \models \varphi]$ THEN $\varphi \in f(P)$.

I.e., if all individuals agree on accepting φ, so also does the collective set.

Responsive (RES) iff $\forall \varphi \in A, \exists P, P' \in \mathbf{P}$ s.t. $\varphi \in f(P)$ AND $\neg \varphi \in f(P')$.

I.e., any formula can possibly be collectively accepted (or rejected) in some profile.

Dictatorial (D) iff $\exists i \in N$ s.t. $\forall P \in \mathbf{P} : f(P) = P_i$.

I.e., there exists an individual (the dictator) whose judgment set is always identical to the collective set.

Oligarchic (O) iff $\exists O \subseteq N$ s.t. $O \neq \emptyset, \forall \varphi \in A, \forall P \in \mathbf{P} : [\bigcap_{i \in O} P_i =_\varphi f(P)]$.

I.e., there exists a non-empty set (the oligarchy) of individuals (oligarchs) s.t. any formula is collectively accepted if and only if it is accepted by each oligarch.

Monotonic (MON) iff $\forall \varphi \in A, \forall i \in N, \forall P, P' \in \mathbf{P}$: IF $[P =_{-i} P'$ AND $P_i \not\models \varphi$ AND $P'_i \models \varphi]$ THEN $[$ IF $\varphi \in f(P)$ THEN $\varphi \in f(P')]$.

I.e., if the collective judgment accepts a formula, then letting one of the individuals that rejects that formula switch to acceptance does not modify the collective judgment.

Independent (IND) iff $\forall \varphi \in A, \forall P, P' \in \mathbf{P}$: IF $[\forall i \in N : P_i =_\varphi P'_i]$ THEN $f(P) =_\varphi f(P')$.

I.e., if all individuals in two different profiles agree on the acceptance or rejection of some formula in the agenda, then the aggregated judgments of the two profiles also agree on the acceptance or rejection of the formula.

Neutral (NEU) iff $\forall \varphi, \psi \in A, \forall P \in \mathbf{P}$: IF $[\forall i \in N : P_i \models \varphi$ IFF $P_i \models \psi]$ THEN $[\varphi \in f(P)$ IFF $\psi \in f(P)]$.

I.e., if all individuals in one same profile accept a formula φ if and only if they accept a formula ψ, then in the aggregated profile φ is accepted if and only ψ is.

Systematic (SYS) iff $\forall \varphi, \psi \in A, \forall P, P' \in \mathbf{P}$: IF $[\forall i \in N : P_i \models \varphi$ IFF $P'_i \models \psi]$ THEN $[\varphi \in f(P)$ IFF $\psi \in f(P')]$.

I.e., if all individuals in two different profiles agree on the acceptance or rejection pattern of two formulae (φ is accepted iff ψ is accepted), the aggregated judgments of the two profiles also do.

Unbiasedness (UNB) iff $\forall \varphi \in A, \forall P, P' \in \mathbf{P}$: IF $[\forall i \in N : P_i \models \varphi$ IFF $P'_i \models \neg \varphi]$ THEN $[\varphi \in f(P)$ IFF $\neg \varphi \in f(P')]$.

I.e., if in two different profiles an individual accepts φ in the first iff it rejects φ in the second, then the aggregation on the first profile accepts φ iff the aggregation on the second rejects it.

[23]Let $\mu : N \longrightarrow N$ be a bijection. The permutation $\mu(P)$ of $P = \langle J_i \rangle_{1 \leq i \leq |N|}$ is the profile $\langle J_{\mu(i)} \rangle_{1 \leq i \leq |N|}$.

The above conditions state some very diverse constraints on how the aggregation maps the input—a judgment profile—to the output—a judgment set. All of them, as we will see in the next section, appeal to some intuition of what counts as a 'fair' or 'reasonable' aggregation process. The theory of judgment aggregation starts by the realization that natural combinations of these conditions lead to unacceptable consequences.

2.3.2   ON THE 'MEANING' OF THE AGGREGATION CONDITIONS

Output conditions

Let us start with the output conditions. Condition **RAT** imposes the result of the aggregation to be a set of formulae of the same type of the ones held by the individual judgments. More loosely, it forces the view of the collective to be like the one of each individual in the group. This involves being consistent, i.e., non-contradictory, and complete, i.e., accepting or rejecting each single issue. Closure is a weaker property than completeness, as it just requires the collective judgment to explicitly accept all the consequences of the formulae it accepts.[24]

Mapping conditions

Let us move then to the mapping conditions. Condition **AN** states that the aggregation is independent of the order in which the individuals' judgment sets appear in profiles. In other words, profiles are treated as multi-sets. Condition **U** simply states that if all individuals agree on the acceptance or rejection of one issue, the aggregated profile agrees too. Condition **RES** imposes that for any formula there is some profile that, once aggregated, accepts that formula. In other words, all formulae have a chance of being collectively accepted. Finally, **D** states that there exists one individual—the *dictator*—and **O** that there exists a set of individuals—the *oligarchy*—who dictate the outcome of every possible aggregation. Clearly, if a function satisfies **AN** it cannot neither be dictatorial nor oligarchic.

The remaining properties are slightly more involved. A good way to illustrate them is by picturing profiles as matrices:

Remark 2.19   Matrix representation of profiles   We have already noted that each judgment set can be thought of as a valuation $J : A \longrightarrow \{1, 0\}$. Given an enumeration of the elements of A, a judgment set can therefore be represented as a tuple (a vector) $\langle J(\varphi_1), \ldots, J(\varphi_{|A|})\rangle$. And given an enumeration of the judgment sets, a profile can be represented as $\langle P_1, \ldots, P_{|N|}\rangle$, and therefore as a matrix where each row is a judgment set, and each column encodes the attitude of each individual toward a formula:

$$\begin{pmatrix} P_1(\varphi_1) & P_1(\varphi_2) & \ldots & P_1(\varphi_{|A|}) \\ P_2(\varphi_1) & P_2(\varphi_2) & \ldots & P_2(\varphi_{|A|}) \\ \vdots & \vdots & \ddots & \vdots \\ P_{|N|}(\varphi_1) & P_{|N|}(\varphi_2) & \ldots & P_{|N|}(\varphi_{|A|}) \end{pmatrix}$$

---

[24]E.g., ∅ satisfies consistency and closure, but not completeness.

where $P_i(\varphi_j)$ is the value (1 for acceptance, 0 for rejection) that judgment set P_i attributes to formula φ_j. That is, each cell represents whether individual i (i.e., the ith row) accepts formula j (i.e., the jth column).[25]

To proceed with our illustration of the meaning of aggregation conditions, consider now two profiles P and P', together with their aggregated sets:

$$
\begin{array}{cccc}
P_1(\varphi_1) & P_1(\varphi_2) & \dots & P_1(\varphi_{|A|}) \\
P_2(\varphi_1) & P_2(\varphi_2) & \dots & P_2(\varphi_{|A|}) \\
\vdots & \vdots & \ddots & \vdots \\
\hline
P_{|N|}(\varphi_1) & P_{|N|}(\varphi_2) & \dots & P_{|N|}(\varphi_{|A|}) \\
\hline
f(P)(\varphi_1) & f(P)(\varphi_2) & \dots & f(P)(\varphi_{|A|})
\end{array}
\quad \text{and} \quad
\begin{array}{cccc}
P'_1(\varphi_1) & P'_1(\varphi_2) & \dots & P'_1(\varphi_{|A|}) \\
P'_2(\varphi_1) & P'_2(\varphi_2) & \dots & P'_2(\varphi_{|A|}) \\
\vdots & \vdots & \ddots & \vdots \\
\hline
P'_{|N|}(\varphi_1) & P'_{|N|}(\varphi_2) & \dots & P'_{|N|}(\varphi_{|A|}) \\
\hline
f(P')(\varphi_1) & f(P')(\varphi_2) & \dots & f(P')(\varphi_{|A|})
\end{array}
$$

where $f(P)(\varphi_j)$—respectively, $f(P')(\varphi_j)$—is the value that the collective set attributes to formula φ_j.

With this in mind, it becomes easier to picture the behavior that each condition imposes on the aggregation. Let us start with **MON**. It says that if all rows in P and P' are identical except for one, say row i, and that row assigns 0 to φ in P and 1 to it in P', then if $f(P)$ assigns 1 so does $f(P')$. Then, **IND** states that, for j s.t. $1 \le j \le |A|$, if the jth column in P consists of the same sequence of zeros and ones as the j-th column in P', then the j-th element in the aggregated sets is the same. Property **NEU** states something similar about two columns in one profile: for any two columns $1 \le j \ne k \le |A|$ in a given profile P, if the jth column consists of the same sequence of zeros and ones as k-th column, then the jth and k^{th} entries in the vector of the collective set are the same.

Finally, property **SYS** pulls **IND** and **NEU** together. It states that, for any two columns $1 \le j \ne k \le m$ in, respectively, profile P and profile P', if the jth column in P consists of the same sequence of zeros and ones as the kth column in P', then the jth and k^{th} entries in the vector of the collective set are the same.[26] So **SYS** is equivalent to the conjunction of **IND** and **NEU**. Finally, **UNB** weakens **SYS** by stating that if the column of φ in P consists of the same sequence of zeros and ones as the column of $\neg\varphi$ in P', then the value assigned to φ by $f(P)$ is the same as the value assigned to $\neg\varphi$ by $f(P')$.

Example 2.20 Properties of aggregation rules   Before concluding this chapter, let us briefly test some of the aggregation rules introduced in Section 2.1.4 against some of the above conditions.

---

[25]Matrices just make the tabular representations of judgment sets we have been familiar with since Chapter 1 (e.g., Figures 1.4 and 1.5) more rigorous. The matrix representation of profiles will come in handy again in Chapter 7.

[26] If f is systematic, then the only information which, in a given profile P, is used by f are properties of the columns of P (e.g., the proportion of 1s and 0s in the column). More precisely, there exists a function g which, for each $1 - 0$ matrix generated by a profile, associates a sequence of $1 - 0$ values such that, for $1 \le j \le m$ (cf. [PvH06]): $f(P_1, \dots, P_n)(\varphi_j) = g(P_1(\varphi_j), \dots, P_n(\varphi_j))$.

Clearly, none of those rules satisfy **D** and they all satisfy **MON**. Rules f_u is not complete and f_{maj} is not consistent—as we have seen with the discursive dilemma and the doctrinal paradox—and hence they do not satisfy **RAT**. On the other hand, they both satisfy **SYS**.[27] As to premise- and conclusion-based rules: f_{pb} satisfies **RAT** and f_{cb} is consistent but not complete (and hence does not satisfy **RAT**). None of them, however, satisfies **NEU** given the distinct role that premises and conclusions play in the aggregation process. The reader is invited to test other combinations of properties on those rules.

2.4   FURTHER TOPICS

The chapter has introduced the nuts and bolts of judgment aggregation as framed within propositional logic, which will be our working framework throughout the book. This framework was first introduced by [LP02] and later developed Dietrich and List in a long series of papers, of which [DL13a] is the most recent contribution. However, this is by no means the only existing framework in which to mathematize judgment aggregation problema. In this final section we point the reader to two important alternatives to the propositional logic framework: the *abstract aggregation* framework and the *general logics* framework. They both generalize the propositional logic framework, albeit in two different ways. While the first one treats the problem of aggregation abstracting away from any concrete logical language, the second one generalizes it from propositional logic to any logic satisfying some basic properties.

2.4.1   ABSTRACT AGGREGATION

As we have seen above while illustrating some of the aggregation conditions, judgment sets can be viewed as functions $J : A \longrightarrow \{1, 0\}$ preserving the meaning of propositional connectives, where the values 1 and 0 stand for *acceptance* and, respectively, *rejection*. Bearing on this view, judgment sets are indeed usually represented as rows of yes/no or 1-0 values, one for each of the issues the agenda is built on. If the agenda A is closed under atoms, i.e., it contains all the atomic propositions occurring in their formulae, then each judgment set J corresponds to a propositional valuation function $J : At \longrightarrow \{1, 0\}$ from those atomic propositions to $\{1, 0\}$.[28]

From this vantage point, one can easily forget the nature of the agenda and simply focus on vectors of 1-0 values, appropriately constrained in order to eliminate 'irrational' vectors, and study aggregation functions as objects of type $f : V^{|N|} \longrightarrow \{1, 0\}^m$ where $V \subseteq \{1, 0\}^m$ is the set of 'rational' vectors and m is the number of issues. This setup, first proposed in social choice theory by [Wil75] as a generalization of preference aggregation, has been widely exploited in judgment aggregation. Key contributions are [DH10a, DH10b] and [NP10a] which developed abstract variants of many of the results we are going to present in the remaining chapters.

---

[27]An extensive axiomatic study of f_{maj} follows in the next chapter.

[28]Many of the examples we will be handling in the book (cf. the doctrinal paradox or Example 2.6) are of this type.

Recently, the abstract framework has been object of further study in the field of artificial intelligence where [GE13] has investigated the interaction between output and mapping conditions by looking at which output conditions would be preserved, from individual to collective judgment sets, under the assumption that given mapping conditions are satisfied by the aggregation function.

2.4.2   GENERAL LOGICS

The paradoxes of the aggregation of judgments we have introduced in Chapter 1 are not a peculiarity of propositional logic, and one can show that similar issues would hamper aggregation problems framed in other logics like many-valued [Got07] or modal [BdV01] logics. This is, in a nutshell, the upshot of the work presented in [Die07]. Aggregation is potentially difficult whenever the to-be-aggregated issues are related by some notion—not necessarily classical—of inconsistency, and hence of logical consequence. The centrality of the notion of logical consequence might have already become apparent to the reader in Section 2.2, where all agenda conditions we considered have been using solely that notion.

In the Tarskian tradition logical consequence can be studied from a structural standpoint as a relation (or operation) satisfying some precise constraints [Tar83].[29] As different logics would define consequence relations obeying different constraints, [Die07] formulates and studies judgment aggregation abstracting away from the specifics of any logic and focusing solely on the type of their consequence relation.

---

[29]For $\models\ \subseteq\ \wp(\mathcal{L}) \times \mathcal{L}$ denoting a consequence relation over language $\mathcal{L}$, the classical constraints (corresponding to a Tarskian closure operator) are: $\varphi \in X$ implies $X \models \varphi$ (extensiveness); $X \models \varphi$ and $X \subseteq Y$ implies $Y \models \varphi$ (monotonicity); $X \models \varphi$ and $\forall \psi \in X : Y \models \psi$ implies $Y \models \varphi$ (idempotency).

CHAPTER 3

Impossibility

Are the discursive dilemma and the doctrinal paradox just quirks of propositionwise majority voting? Or would any other seemingly reasonable aggregation procedure run into similar troubles? In this chapter we show how these questions can be answered from a very general standpoint, using the so-called *axiomatic method*. Aggregation conditions are chosen as axioms in an attempt to restrict the set of possible aggregation procedures to a few desirable ones. The axiomatic method is the most influential methodological tool of social choice theory since Arrow's theorem [Arr50], and the chapter showcases its application to judgment aggregation.

Chapter outline: The axiomatic method is first illustrated in Section 3.1, where the proposition-wise majority rule is proven to be, on simple agendas, the only aggregation function that satisfies some highly desirable aggregation conditions. Section 3.2 moves to richer agendas and proves one of the many so-called *impossibility theorems* of judgment aggregation, showing that rather undemanding conditions on the aggregation function force the aggregation to be dictatorial, i.e., such that one individual always decides the outcome of the aggregation. The theorem, due to Dietrich and List [DL07a], will be proven by using a widespread proof technique in the literature on social choice theory: the ultrafilter proof technique. Section 3.3 discusses the result and its proof in further detail and, using the same technique, proves a related impossibility result [DL08] based on the existence of an oligarchy rather than a dictatorship. Finally, Section 3.4 discusses and provides pointers to other similar results in the literature, including the case of infinite electorates, and touches upon the relationship between preference and judgment aggregation. The chapter builds and elaborates on material taken mainly from [DL07a, DL08, KE09, Lis12] and [Odi00].

3.1   WHAT IS THE MAJORITY RULE LIKE?

The previous chapter has introduced formal definitions of properties of agendas and aggregation functions. Everything is in place to study what happens of judgment aggregation when some of those properties (or their negations) are imposed as axioms upon the aggregation. Ultimately, this will lead us to unveil the incompatibility of many natural bundles of such properties, through the so-called impossibility theorems, to which we will turn in much detail in Section 3.2.

First, however, we will concern ourselves with an example of a somewhat more 'positive' type of result. The reader might have noticed that, among the aggregation functions introduced at the end of Chapter 1, propositionwise majority is the one we have encountered most often in the examples discussed so far. The reason is that propositionwise majority voting can be shown not only to enjoy a number of desirable properties, but also that, on a specific class of agendas, it is also the only function satisfying a bundle of those desirable properties.

3.1.1   PROPERTIES OF PROPOSITIONWISE MAJORITY

There are good reasons for putting special emphasis on propositionwise majority:

Fact 3.1   Let $\mathcal{J} = \langle N, A \rangle$ be a judgment aggregation problem:

i) f_{maj} does not satisfy **D**;

ii) f_{maj} satisfies **U, RES, AN, MON, IND, NEU, SYS, UNB**;

iii) If A is simple, then f_{maj} satisfies **RAT** iff $|N|$ is odd.

Sketch of proof. In the case of f_{maj}, in each profile P, the only information needed to decide whether a formula φ is collectively accepted is the integer $|P_\varphi|$, i.e., the number of individuals accepting φ. Given this observation, claims i) and ii) follow fairly straightforwardly and are left to the reader. We focus on claim iii). $\boxed{\Leftarrow}$ We show that if A is simple and $|N|$ is odd then f_{maj} is: (a) consistent and (b) complete. (a) If A is simple then each minimally inconsistent set X is such that $|X| < 3$ (Definition 2.7), viz. $X = \{\varphi, \psi\}$ for $\varphi \models \neg\psi$. Suppose toward a contradiction that for some P, $f_{maj}(P)$ is inconsistent. It must then contain a minimally inconsistent set $\{\varphi, \psi\}$. So, by the definition of f_{maj}, both $|P_\varphi|$ and $|P_\psi|$ are greater or equal to $\lceil \frac{N+1}{2} \rceil$. We have two cases: (a') $P_\varphi \cap P_\psi = \emptyset$; (a") $P_\varphi \cap P_\psi \neq \emptyset$. If (a'), then $|P_\varphi| + |P_\psi| > N$, which is impossible. If (a"), then for some $i \in N$, $P_i \models \varphi$ and $P_i \models \psi$, which is impossible by the definitions of aggregation function (Definition 2.4) and judgment set (Definition 2.2). Contradiction. (b) If $|N|$ is odd then f_{maj} clearly outputs a complete set of judgments as for any issue φ there will either be a majority for φ or for $\neg\varphi$. $\boxed{\Rightarrow}$ We show that if A is simple and f_{maj} satisfies **RAT** then $|N|$ is odd. Assume, toward a contradiction, that $|N|$ is even. Take now a profile P such that $|P_\varphi| = |P_{\neg\varphi}|$. We have that $\varphi \notin f_{maj}(P)$ and $\neg\varphi \notin f_{maj}(P)$, against the assumption of **RAT**. □

So, the majority rule implements all among the desirable mapping conditions (Definition 2.18) we mentioned in the previous chapter, and when considering the case of simple agendas, it also guarantees the collective judgment to be consistent and complete (when the number of voters is odd). The natural question is then: are there other aggregation functions with the same features? The answer is no, as we proceed now to show.

3.1.2   CHARACTERIZING PROPOSITIONWISE MAJORITY

When aggregating simple agendas, there are no 'reasonable' aggregation functions other than the propositionwise majority rule:

Theorem 3.2   Characterization of f_{maj}**.**   *Let* $\mathcal{J} = \langle N, A \rangle$ *be a judgment aggregation problem where A is simple and* $|N|$ *is odd: an aggregation function f for* $\mathcal{J}$ *is the propositionwise majority rule* f_{maj} *if and only if it satisfies* **RAT**, **AN**, **MON** *and* **UNB**.

Proof. $\boxed{\Rightarrow}$ The claim follows directly from Fact 3.1. $\boxed{\Leftarrow}$ Assume f satisfies **AN**. Then for each profile P and formula $\varphi \in A$, the only information that f uses to determine whether $\varphi \in f(P)$ is $|P_\varphi|$. That is, P can be abstracted to $\langle |P_\varphi| \rangle_{\varphi \in A}$. We proceed by case distinction: (i) Assume that, for any $\varphi \in A$ and for any $P \in \mathbf{P}$ s.t. $|P_\varphi| = |P_{\neg\varphi}| + 1$, $\varphi \in f(P)$. If this is the case, by **MON** we have that $\varphi \in f(P)$ iff $|P_\varphi| > |P_{\neg\varphi}|$ or, equivalently, $|P_\varphi| \geq \left\lceil \frac{|N|+1}{2} \right\rceil$. Therefore f is the propositionwise majority rule (Formula 2.1). (ii) Assume that, for some $\varphi \in A$ and $P \in \mathbf{P}$, $|P_\varphi| = |P_{\neg\varphi}| + 1$ but $\varphi \notin f(P)$. By **RAT** we have that $\neg\varphi \in f(P)$. Now consider a profile P' obtained from P by letting an individual in P_φ accept $\neg\varphi$ instead. By **MON** we still have that $\neg\varphi \in f(P')$. We also have that $|P'_{\neg\varphi}| = |P'_\varphi| + 1$, from which by **UNB** we obtain that $\varphi \in f(P')$. Contradiction. $\square$

The theorem can rightly be seen as a judgment aggregation variant of May's theorem, the well-known characterization of majority voting in preference aggregation [May52].[1]

Voting on simple agendas involves only the choice between collectively accepting or rejecting issues with limited logical interdependencies. This is not an uncommon setting. For instance, multilateral treaties among states often take this form with individual states having to vote on whether a given provision is to be incorporated or not in a treaty [BKS07]. The import of the theorem is that, in such a context (and with an odd number of voters), propositionwise majority is the only aggregation function that is collectively rational, anonymous, monotonic and unbiased. In other words, a 'reasonable' voting procedure exists and it is moreover unique. The situation changes drastically when we move to richer agendas.

3.2   AN IMPOSSIBILITY THEOREM

Impossibility theorems are the main type of results that have driven the literature on preference and judgment aggregation (cf. the historical discussion of Chapter 1). Their gist consists in showing that imposing seemingly desirable constraints on the aggregation problem—i.e., on the

---

[1]To the best of our knowledge, the characterization of propositionwise majority of Theorem 3.2 is novel. Other characterization results can be found in the literature. In particular, [DL10b] proves a stronger theorem, with a slightly more involved proof. It drops the assumption that $|N|$ be odd, and shows that propositionwise majority is the only aggregation function which satisfies anonymity, a weaker form of unbiasedness known as acceptance/rejection neutrality, and which is collectively consistent (but not necessarily complete).

agenda and on the aggregation function—may lead to degenerate forms of aggregation, typically, to dictatorships.

The present section is devoted to the statement and proof—through the so-called *ultrafilter technique*—of one representative impossibility theorem on the aggregation of judgments. Due care will be taken in explaining and illustrating the ultrafilter proof technique, which is arguably one of the most important tools in the toolbox of an 'aggregation theorist', and which has been applied to obtain a variety of impossibility theorems in both preference and judgment aggregation. So let us state the theorem first, which is due to [DL07a]:

> Let the agenda be non-simple and even number negatable: an aggregation function is collectively rational, unanimous and systematic if and only if it is a dictatorship by some individual.

Or put otherwise, it is impossible to aggregate in a non-trivial way—like dictatorship does—individual judgments sets into a collective judgment set by respecting unanimity and systematicity. The reader will find the theorem restated at the end of the section as Theorem 3.7.

We now turn to its proof, for which we will need three lemmas. The first one relates the condition of systematicity to the possibility of defining a set of coalitions of voters which can force the whole collective judgment. The second one makes explicit the specific structure of this set of coalitions. Finally, the third one establishes the existence of a dictator.

3.2.1   WINNING COALITIONS

Given any judgment aggregation problem and aggregation function, we can ask ourselves for which agents it always holds that if they all at the same time accept a given formula, so does the collective judgment. In other words, we can always define for any element φ of the agenda, what the coalition is of agents that can always force φ to be collectively accepted. Such coalitions are called decisive or *winning*.

Coalitions that are winning for φ

Definition 3.3   Winning coalitions for φ.   Let $\mathcal{J} = \langle N, A \rangle$ be a judgment aggregation problem, f an aggregation function and $\varphi \in A$. A coalition $C \subseteq N$ is *winning for φ* iff:

$$\forall P \in \mathbf{P} : \text{ IF } C = P_\varphi \text{ THEN } \varphi \in f(P).$$

The set of winning coalitions for φ in $\mathcal{J}$ under f is denoted $\mathcal{W}_\varphi(\mathcal{J}, f)$.[2]

Intuitively, a set of individuals is winning for an element of the agenda if, pulling their votes together, they can guarantee that element to be collectively accepted.

[2]We will usually drop the reference to $\mathcal{J}$ and f as they will usually be clear from the context.

One could already observe the relevance of aggregation conditions such as **IND** and **SYS** for the existence of winning coalitions. By both those conditions, if there exists a profile P such that $C = \{i \in N \mid P_i \models \varphi\}$ and $\varphi \in f(P)$, then for all P' such that $C = \{i \in N \mid P_i' \models \varphi\}$ it holds that $\varphi \in f(P')$. In other words, if C is winning for φ *in one profile*, then it is winning for φ in all profiles, that is, it is a winning coalition for φ.

All-winning coalitions

On the same line, consider now the following set, which contains all those coalitions that are *winning for all* formulae in the agenda:

$$\mathcal{W} := \{C \subseteq N \mid \forall \varphi \in A, \forall P \in \mathbf{P} : \text{IF } C = P_\varphi \text{ THEN } \varphi \in f(P)\} \tag{3.1}$$

We will refer to $\mathcal{W}$ simply as the set of winning coalitions.

It turns out that an aggregation function is systematic if and only if it can be described by a set of winning coalitions, in the following sense: if f is systematic then it will accept φ if and only if the set of voters accepting φ in the profile is a member of $\mathcal{W}$, and *vice versa*.

Lemma 3.4 Characterization of SYS by $\mathcal{W}$ *Let $\mathcal{J} = \langle N, A \rangle$ be a judgment aggregation problem and f an aggregation function. These two statements are equivalent:*

   *i) f satisfies **SYS**;*

   ii) $\varphi \in f(P)$ iff $P_\varphi \in \mathcal{W}$, for all $P \in \mathbf{P}$ and $\varphi \in A$.

Proof. $\boxed{\text{From i) to ii)}}$ Assume i) is the case and prove ii). $\boxed{\Leftarrow}$ The claim holds directly by the above definition of $\mathcal{W}$ (Formula 3.1). $\boxed{\Rightarrow}$ Assume $\varphi \in f(P)$ and consider the set of voters P_φ. For any $P' \in \mathbf{P}$, by **SYS** we have that if $P_\varphi = P_\varphi'$ then $\varphi \in f(P')$. Hence $P_\varphi \in \mathcal{W}$ according to the definition of $\mathcal{W}$ in Formula 3.1. $\boxed{\text{From ii) to i)}}$ Assume ii) is the case and, toward a contradiction, that f does not satisfy **SYS**, that is, $\exists \varphi, \psi \in A$ and $\exists P, P' \in \mathbf{P}$ s.t. $P_\varphi = P_\psi'$ and $\varphi \in f(P)$ and $\psi \notin f(P')$. By ii) it follows that $P_\varphi = P_\psi' \in \mathcal{W}$ and thus $\psi \in f(P')$. Contradiction. $\qquad\square$

Notice that the result goes through independently of assumptions on the richness of the agenda and, noticeably, on the rationality of the collective set resulting from the aggregation.

3.2.2   WINNING COALITIONS AS ULTRAFILTERS

We now move to the central lemma in the proof. Under some conditions on the complexity of the agenda—even-negatability in our case—if the aggregation function is unanimous, systematic and collectively rational, then the set of winning coalitions $\mathcal{W}$ takes the form of an ultrafilter. Ultrafilters were originally introduced in [Car37] to capture a handful of properties characterizing the notion of 'large set', like: i) the largest set is a large set; ii) a set is large iff its complement is not large; iii) if a set is large its supersets are also large; iv) the intersection of two large sets is large.

The intuition behind the use of ultrafilters in voting theory is that a notion of 'large set' can naturally be used to define independent aggregation procedures responding to the rough intuition: issue φ is collectively accepted if and only if there is a large set of individuals—a large coalition—supporting it. So the upshot of the following lemma is that, under the stated conditions of the theorem, the aggregation function is indeed characterized by a set of winning coalitions where being 'winning' means being 'large' in the sense captured by ultrafilters.

Lemma 3.5   Ultrafilter lemma   *Let $\mathcal{J} = \langle N, A \rangle$ be a judgment aggregation problem and f an aggregation function such that A satisfies NS and EN and f satisfies* **U**, **SYS** *and* **RAT**. *The set $\mathcal{W}$ is an* ultrafilter, *i.e.:*

i) *$N \in \mathcal{W}$, i.e., the set of all individuals is a winning coalition;*

ii) *$C \in \mathcal{W}$ iff $-C \notin \mathcal{W}$, i.e., a coalition is winning if and only if its complement is losing;*

iii) *$\mathcal{W}$ is upward closed: if $C \in \mathcal{W}$ and $C \subseteq C'$ then $C' \in \mathcal{W}$, i.e., if a coalition is winning, all coalitions containing it are also winning;*

iv) *$\mathcal{W}$ is closed under finite intersections: if $C, C' \in \mathcal{W}$ then $C \cap C' \in \mathcal{W}$, i.e., if two coalitions are winning then the individuals they have in common form a winning coalition.*

Proof. Proofs follow for each of the four claims:

i) The claim is a direct consequence of the assumption that f satisfies **U**.

ii) $\boxed{\Rightarrow}$ Suppose, toward a contradiction, that $C, -C \in \mathcal{W}$. Consider now a profile where the judgment sets of the agents in C contain φ and those in $-C$ contain $\neg\varphi$. This profile must exist by the definition of aggregation function (Definition 2.4), and it would be inconsistent, which is impossible by the assumption of **RAT**. $\boxed{\Leftarrow}$ By contraposition, suppose $C \notin \mathcal{W}$. Then, by Lemma 3.4, $\forall \varphi \in A, P \in \mathbf{P}$ we have that if $P_\varphi = C$ then $\varphi \notin f(P)$ and therefore by **RAT** that $\neg\varphi f(P)$. Since judgment sets are complete, this is equivalent to stating that $\forall \varphi \in A, P \in \mathbf{P}$ if $P_{\neg\varphi} = -C$ then $\neg\varphi \in f(P)$. Hence, $-C \in \mathcal{W}$ (Formula 3.1).

iii) We proceed toward a contradiction: assume $C \in \mathcal{W}$, $C \subseteq C'$ and $C' \notin \mathcal{W}$. Take a minimally inconsistent set $X \subseteq A$ s.t. $\exists Y \subset X$ with $Y = \{\varphi, \psi\}$ for $\varphi, \psi \in A$ and s.t. $(X - Y) \cup \{\neg\varphi, \neg\psi\}$ is consistent. This set exists by EN (Definition 2.10). Since X is minimally inconsistent, it follows that $(X - \{\varphi\}) \cup \{\neg\varphi\}$ and $(X - \{\psi\}) \cup \{\neg\psi\}$ are consistent. Consistent is also, by the definition of EN, the set $(X - \{\varphi, \psi\}) \cup \{\neg\varphi, \neg\psi\}$. Consider now these three

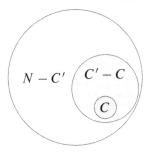

Figure 3.1: Venn diagram of the tripartition of the set of voters defined in the proof of claim iii) Lemma 3.5.

coalitions which, notice, form a partition of N in three subsets (see Figure 3.1 for a depiction of this coalitional structure):

$$
\begin{aligned}
C_1 &:= C \\
C_2 &:= C' - C \\
C_3 &:= N - C'
\end{aligned}
$$

and consider the judgment profile P defined as follows (double negations are removed):[3]

$$
P_i = \begin{cases}
(X - \{\varphi\}) \cup \{\neg\varphi\} & \text{IF } i \in C_1 \\
(X - \{\varphi, \psi\}) \cup \{\neg\varphi, \neg\psi\} & \text{IF } i \in C_2 \\
(X - \{\psi\}) \cup \{\neg\psi\} & \text{IF } i \in C_3
\end{cases}
$$

We can now conclude the following about P. By **U**, we have that $N \in \mathcal{W}$ and hence $X - \{\varphi, \psi\} \subseteq f(P)$. Since $C \in \mathcal{W}$ by assumption, we also have that $\psi \in f(P)$. Furthermore, by item ii) in this lemma, and the assumption that $C' \notin \mathcal{W}$ we conclude that $C_3 \in \mathcal{W}$ and consequently that $\varphi \in f(P)$. It follows that $X \subseteq f(P)$ where X was assumed to be inconsistent, against the assumption that f satisfies **RAT**.

iv) Assume toward a contradiction that $C, C' \in \mathcal{W}$ and $C \cap C' \notin \mathcal{W}$. By NS there exists a minimally inconsistent set $X \subseteq A$ s.t. $3 \le |X|$ (Definition 2.7). Take three elements of X: φ, ψ, ξ. By the same definition we have that for $x \in \{\varphi, \psi, \xi\}$: $(X - \{x\}) \cup \{\neg x\}$ is consistent. Consider now these three coalitions which, notice, form a partition of N (see Figure 3.2 for a

---

[3]Let us also give a concrete example of the construction for the agenda $A = \pm\{p, q, p \wedge q\}$. The minimally inconsistent set is $X = \{p, q, \neg(p \wedge q)\}$ and the profile is:

$$
P_i = \begin{cases}
\{q, \neg(p \wedge q), \neg p\} & \text{IF } i \in C_1 \\
\{\neg p, \neg(p \wedge q), \neg q\} & \text{IF } i \in C_2 \\
\{p, \neg(p \wedge q), \neg q\} & \text{IF } i \in C_3
\end{cases}
$$

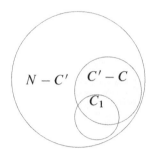

Figure 3.2: Venn diagram of the tripartition of the set of voters defined in the proof of claim iv) Lemma 3.5.

depiction of this coalitional structure):

$$
\begin{aligned}
C_1 &:= C \cap C' \\
C_2 &:= C' - C \\
C_3 &:= N - C'
\end{aligned}
$$

and consider the judgment profile P defined as follows (double negations are removed):[4]

$$
P_i = \begin{cases}
(X - \{\varphi\}) \cup \{\neg\varphi\} & \text{if } i \in C_1 \\
(X - \{\xi\}) \cup \{\neg\xi\} & \text{if } i \in C_2 \\
(X - \{\psi\}) \cup \{\neg\psi\} & \text{if } i \in C_3
\end{cases}
$$

By **U** we have that $X - \{\varphi, \psi, \xi\} \subseteq f(P)$. Since $C' = C_1 \cup C_2 \in \mathcal{W}$, it follows that $\psi \in f(P)$. Also, since $C \subseteq C_1 \cup C_3$, by claim iii) above we have that $C_1 \cup C_3 \in \mathcal{W}$. Hence $\xi \in f(P)$. Finally, since by assumption $C \cap C' = C_1 \notin \mathcal{W}$, by claim ii) we have that $C_2 \cup C_3 \in \mathcal{W}$ and hence that $\varphi \in f(P)$. From this we conclude that $X \subseteq f(X)$. But X was assumed to be inconsistent, so the conclusion contradicts condition **RAT**.

This completes the proof.[5]   □

We will comment further on the use of ultrafilters in social choice theory and judgment aggregation later in Section 3.3.3.

---

[4]This is a concrete example for the discursive dilemma agenda $A = \pm\{p, q, p \wedge q\}$. The minimally inconsistent set is, again, $X = \{p, q, \neg(p \wedge q)\}$ and the profile is:

$$
P_i = \begin{cases}
\{q, \neg(p \wedge q), \neg p\} & \text{if } i \in C_1 \\
\{p, p \wedge q, q\} & \text{if } i \in C_2 \\
\{p, \neg(p \wedge q), \neg q\} & \text{if } i \in C_3
\end{cases}
$$

So, $\varphi := p, \psi := q, \xi := \neg(p \wedge q)$.

[5]Some readers might have noticed that condition *iii)* in the definition of ultrafilter could be dispensed with, as it follows from the other three: if $C' \notin \mathcal{W}$ then, by *ii)* we obtain that $-C' \in \mathcal{W}$, and by *iv)* that $C \cap -C' = \emptyset \in \mathcal{W}$, which is impossible given *i)* and *ii)*. Nevertheless, it must be noticed that all conditions need to be established in the proof of the ultrafilter lemma. To appreciate this notice that, in order to prove that condition *iv)* holds (using the non-simplicity assumption), one needs to prove that condition *iii)* holds (using the even-number negatability assumption).

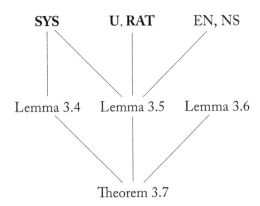

Figure 3.3: Structure of the proof of Theorem 3.7. Edges indicate dependences (from top to bottom) between assumptions, lemmas and the impossibility theorem.

3.2.3   DICTATORS

We conclude with the last lemma needed to establish the theorem. Interestingly, this lemma consists of a property of all finite ultrafilters, i.e., ultrafilters which are defined, like in our case, on a finite domain. In fact, we will see, the proof does not involve any reference to aggregation problems and functions.

Lemma 3.6   Existence of a dictator   *Let $\mathcal{W}$ be an ultrafilter on a finite set N. Then $\mathcal{W}$ is* principal, *i.e.: $\exists i \in N$ s.t. $\{i\} \in \mathcal{W}$.*

Proof. Recall the definition of ultrafilter given within Lemma 3.5. We will be using the properties *i)*, *ii)* and *iv)* of ultrafilters as defined there. Since $|N|$ is finite, the closure $\bigcap \mathcal{W}$ of $\mathcal{W}$ under finite intersections belongs to $\mathcal{W}$ by property *iv)*. We therefore have that $\bigcap \mathcal{W} \neq \emptyset$. For suppose not, then $N \notin \mathcal{W}$ by property *ii)*, against property *i)*. So, W.L.O.G. , assume $i \in \bigcap \mathcal{W}$ for $i \in N$. Suppose toward a contradiction that $\{i\} \notin \mathcal{W}$. By property *ii)* we have that $N - \{i\} \in \mathcal{W}$, from which follows that $i \notin \bigcap \mathcal{W}$. Contradiction. Hence $\{i\} \in \mathcal{W}$.   □

Now, since we have showed that the set of winning coalitions is a (finite) ultrafilter, this lemma tells us that in each such set of winning coalitions there always exists a voter who is a winning coalition. Such voter is therefore a (unique) dictator. It is worth stressing again that this lemma does not hinge on any specific judgment aggregation property or construction, but expresses a general property of finite ultrafilters.

3.2.4   THE THEOREM

We can now pull the above lemmas together and prove the result we were after:

Theorem 3.7 [DL07a].   *Let $\mathcal{J} = \langle N, A \rangle$ be a judgment aggregation problem such that A satisfies NS and EN, and let f be an aggregation function: f satisfies* **U**, **RAT** *and* **SYS** *iff f satisfies* **D**.

Proof. $\boxed{\Leftarrow}$ It is easy to verify that if f satisfies **D** then it trivially satisfies **U**, **RAT** and **SYS**. $\boxed{\Rightarrow}$ By Lemma 3.4, for any $P \in \mathbf{P}$ and $\varphi \in A$:

$$\varphi \in f(P) \quad \text{IFF} \quad P_\varphi \in \mathcal{W}.$$

Then, by Lemma 3.5 and 3.6 we have that $\{i\} \in \mathcal{W}$ for some $i \in N$ and hence:

$$P_\varphi \in \mathcal{W} \quad \text{IFF} \quad i \in P_\varphi$$

which concludes the proof: $\varphi \in f(P)$ iff $P_i \models \varphi$. $\square$

So if the agenda exhibits some level of complexity in the logical interrelationships among its elements, requiring the aggregation function to be collectively rational, unanimous and systematic amounts to require the aggregation function to be a dictatorship.

The first impossibility theorem of judgment aggregation, proven in the paper that initiated the field [LP02], is a direct consequence of Theorem 3.7. Agendas such as $\pm\{p, q, p \wedge q\}$ or $\pm\{p, q, p \rightarrow q\}$, which satisfy NS and EN, can be aggregated only in a trivial way, via a dictatorship, if we are to guarantee that the aggregation is unanimous and systematic.

The diagram in Figure 3.3 recapitulates the structure of the proof highlighting the dependences between the five assumptions, the three lemmas, and the final statement.

3.3   (ULTRA)FILTERS, DICTATORS AND OLIGARCHS

Non-dictatorship might be considered as a 'mild' condition to impose on a democratic process of aggregation. Even if no dictator exists, it might still be possible for the aggregation to be fully determined by what a special set of voters agrees upon. More formally, the aggregation function f can be such that there exists a non-empty set of voters O such that: $\forall P \in \mathbf{P} : f(P) = \bigcap_{i \in O} J_i$.

In Chapter 2 we have called aggregation functions for which this is the case *oligarchic*[6] and in this section we study impossibility results related to them.

---

[6]Other notions of oligarchy are discussed in the literature. For example: weak oligarchy [DL08, Gär06], for which $f(J_1, \ldots, J_n) \subseteq \bigcap_{i \in O} J_i$; oligarchies with a default [NP06], for which every time no agreement is found among the oligarchs, the collective judgment is set by a default judgment set.

3.3.1   IMPOSSIBILITY OF NON-OLIGARCHIC AGGREGATION

There are two extreme cases of oligarchies: dictatorships, where O is a singleton; and unanimities, where O equals the set of individuals N. In the first case, the oligarchy generates complete judgment sets, while in the second the oligarchy always generates incomplete judgments except in the rare cases where all individuals agree on all issues.

Neither of the two cases is desirable and the notion of oligarchy is therefore a natural one upon which to base impossibility results analogous to the ones holding for the notion of dictatorship:

If the agenda satisfies constraints $C_1, \ldots, C_n$, then the aggregation function satisfies constraints $C_1', \ldots, C_n'$ if and only if the aggregation function is an oligarchy.

As an illustration of this type of results, we will be discussing here the following theorem:

Theorem 3.8   [DL08].   *Let $\mathcal{J} = \langle N, A \rangle$ be a judgment aggregation problem such that A satisfies NS and EN, and let f be an aggregation function: f is consistent, closed,[7] satisfies U and SYS iff f satisfies O.*

In comparison with Theorem 3.7, the use of condition **O** is accompanied by the weakening of **RAT** to the condition that the collective judgment set be consistent and closed. This is a typical trait of impossibility theorems involving oligarchies, and we will come back to it later in Chapter 4.

3.3.2   PROOF: FROM ULTRAFILTERS TO FILTERS

Theorems like the above can be proven with a strategy similar to the one followed with Theorem 3.7, capitalizing again on the structure of the winning coalitions that characterize the aggregation function, thanks to the assumption of **SYS**.[8] In the case of Theorem 3.8, the relevant structure is that of a (proper) filter, instead of that of an ultrafilter:

Lemma 3.9   Filter lemma   *Let $\mathcal{J} = \langle N, A \rangle$ be a judgment aggregation problem and f an aggregation function such that A satisfies NS and EN and f is consistent, closed and satisfies U and SYS. The set of winning coalitions $\mathcal{W}$ is a proper filter, i.e.:*

i) $N \in \mathcal{W}$, *i.e., the set of all individuals is a winning coalition;*

ii) $\emptyset \notin \mathcal{W}$, *i.e., the empty coalition is not winning;*

iii) $\mathcal{W}$ *is upward closed: if $C \in \mathcal{W}$ and $C \subseteq C'$ then $C' \in \mathcal{W}$, i.e., if a coalition is winning, all coalitions containing it are also winning;*

iv) $\mathcal{W}$ *is closed under finite intersection: if $C, C' \in \mathcal{W}$ then $C \cap C' \in \mathcal{W}$, i.e., if two coalitions are winning then the individuals they have in common form a winning coalition.*

---

[7]Recall Definition 2.17.
[8]In fact, the proof we provide here differs substantially from the one provided in [DL08].

Sketch of proof. The proof is an adaptation of the proof of Lemma 3.5 resorting to the weaker assumption of deductive closure of the collective judgment instead of **RAT**.

i) The claim is a direct consequence of the assumption that f satisfies **U**.

ii) Toward a contradiction, if $\emptyset \in W$ then for any profile P and formula φ, $\varphi \in f(P)$, against the consistency assumption on the output of f.

iii) Assume $C \in W$ and $C \subseteq C'$. We proceed with the construction in item iii) of Lemma 3.5, building the following profile where $C = C_1$ and $C' = C_1 \cup C_2$:

$$
P_i = \begin{cases}
(X - \{\varphi\}) \cup \{\neg\varphi\} & \text{IF } i \in C_1 \\
(X - \{\varphi, \psi\}) \cup \{\neg\varphi, \neg\psi\} & \text{IF } i \in C_2 \\
(X - \{\psi\}) \cup \{\neg\psi\} & \text{IF } i \in C_3
\end{cases}
$$

We show that if $C' \in W$. By item i) we have that $N \in W$ and hence $X - \{\varphi, \psi\} \subseteq f(P)$. Since $C \in W$ we have that $\psi \in f(P)$ and since $f(P)$ is closed by assumption, we have that $\neg\varphi \in f(P)$. From this, the fact that $C' = P_{\neg\varphi}$ and that f satisfies **SYS** we obtain that $C' \in W$.

iv) One proceeds in a similar fashion as in item iv) of Lemma 3.5. The details are left to the reader.

This completes the proof. □

So the set of winning coalitions W behaves in this case like in Lemma 3.9 except for the fact that a coalition belongs to the set if and only if its negation does not belong to it—condition ii) in Lemma 3.9. Observe that the failure of this condition is due to the collective judgment set not being necessarily complete.

To complete the proof of Theorem 3.8, one then has to use a general fact about finite proper filters, in the same way as the fact that every finite ultrafilter needs to be principled (Lemma 3.6) has been used to establish dictatorship:

Lemma 3.10 Existence of oligarchs   *Let W be a proper filter on a finite domain N. Then $\exists O \subseteq N$ s.t. $\forall X \in W : \emptyset \neq O \subseteq X$.*

Proof. Since W is proper it follows that: $\emptyset \neq \bigcap W \in W$. □

Theorem 3.8 then follows directly from Lemmas 3.9 and 3.10. More theorems of this type concerning property **O** can be found in [DL08]. We will briefly come back to some of them in Chapter 4.

3.3.3   IMPOSSIBILITY VIA (ULTRA)FILTERS

The proofs we have provided of Theorems 3.7 and 3.8—and the one we will provide of Theorem 5.12—rely critically on the set of winning coalitions of a function exhibiting a specific structure. The general strategy of such proofs can be summarized as follows:

> To establish impossibility results one shows that the conditions imposed on the agenda and the aggregation function force the set of winning coalitions characterizing the function to be an ultrafilter (resp. a proper filter) on the set of voters. If the set of voters is finite, one can then conclude that the ultrafilter is principled, i.e., it is generated by one single element (resp. one non-empty set of elements) that belongs to all winning coalitions, hence establishing the existence of one dictator (resp. one oligarchy).

The first application of this technique, well-established by now in social choice theory, is due to [Fis70, KS72, Han76], all of which offered alternative proofs of Arrow's theorem. In judgment aggregation, several proofs resort explicitly (e.g., [Gär06, KE09, DM10, Her10]) or more often implicitly ([DL07a] itself) to this technique.

We conclude the discussion of filters and ultrafilters in the context of aggregation by relating them to cooperative game theory and to model theory.

Ultrafilters, aggregation and cooperative games

Ultrafilters are special cases of structures known to game theorists as *simple games*. Simple games (cf. [LBS08, Ch. 8]) are coalitional games where the possible payoffs for coalitions are 1—in which case a coalition is winning—or 0—in which case a coalition is losing. So a simple game is a tuple $G = \langle N, \mathcal{W} \rangle$ where N is a set of agents and $\mathcal{W}$ is the set of winning coalitions. Each voting contest can be conceptualized as a simple game, where winning coalitions are the set of voters that, by agreeing on their ballots, can force the outcome of the election. From the point of view of simple games, impossibility theorems like the ones discussed in this chapter amount to showing that, when voting exhibits some critical properties then the associated simple game is a finite ultrafilter (resp. a finite proper filter), and therefore possesses a winning coalition consisting of just one player (resp. of a non-empty set of players).[9]

Ultrafilters, aggregation and model theory

The application of model theoretic methods to preference aggregation was first put forth in [LvL95], where a representation of preference aggregation functions via a construction known as ultraproducts was first described. Recently such technique has been applied and extended within judgment aggregation in a series of papers by Herzberg [Her08, Her13] and Herzberg and Eckert [HE12, Her12].

---

[9] The reader is referred to [BBM81] for more information about the relationships between simple games and ultrafilters in social choice theory and to [Dan10], which explores the connection between simple games and judgment aggregation from a logical angle.

We sketch here just the key idea underpinning this interesting line of work and refer the reader to the above papers for a comprehensive exposition. We know that if the agenda of an aggregation problem is sufficiently rich (e.g., NS and EN) and the aggregation function f satisfies some specific properties (e.g., **SYS**, **U** and **RAT**) then the function is characterized by an ultrafilter of winning coalitions (e.g. Lemmas 3.4 and 3.5), which we may call here $\mathcal{W}_f$. What this line of work showed is that the application of the aggregation function f to profile P actually corresponds to the construction of the ultraproduct $\prod P/\mathcal{W}_f$ of P, with respect to the ultrafilter $\mathcal{W}_f$, where:[10]

$$\prod P/\mathcal{W}_f \;=\; \{\varphi \in A \mid P_\varphi \in \mathcal{W}_f\} \tag{3.2}$$

Put simply, one obtains a representation theorem for the class of all aggregation functions (with the above properties) of this form: $f(P) = \prod P/\mathcal{W}_f$ for any $P \in \mathbf{P}$. Once this bridge is laid, further properties of aggregation functions can be derived [Her08] (like **D**, in the case of finite sets of voters, or **MON**) as well as new impossibility results [Her12], the aggregation problem can be generalized to any first-order logic theory [HE09], and further model-theoretic representations of aggregation functions become available [Her10].

3.4 FURTHER TOPICS

In this section we wrap up pointing the reader to some more impossibility results strictly related to the ones proven above, to the analysis of infinite electorates, and we conclude with a few more comments on the relationships between preference and judgment aggregation.

3.4.1 OTHER IMPOSSIBILITY RESULTS

Impossibility theorems are negative answers to the following question about the *possibility* of aggregation:[11]

> If the agenda satisfies the agenda conditions $C_1, \ldots, C_n$, does an aggregation function exist, which satisfies the aggregation conditions $C'_1, \ldots, C'_m$ (typically including collective rationality)?

The negative answer is then typically stated in the form of a characterization of dictatorial aggregation functions:

> *If* the agenda satisfies the agenda conditions $C_1, \ldots, C_n$, *then* the aggregation function satisfies the aggregation conditions $C'_1, \ldots, C'_m$ (typically including collective rationality) *if and only if* the aggregation function is a *dictatorship* (or an *oligarchy*).

---

[10]We use here the notion of ultraproduct as defined on a profile of sets, rather than on models, and we refrain from giving the full definition from which Formula 3.2 is obtained. Ultraproducts are a standard technique in model theory to construct new models from collections of old ones—in our case a new judgment set from a profile of old ones. These new models are such that they satisfy a formula if and only if a 'large set' of the old ones do. The interested reader is referred to [Hod97, Ch. 8] for a detailed exposition of this construction.

[11]Arrow himself refers to his theorem in [Arr50, Arr63] as the "General Possibility Theorem."

Agenda conditions	Aggregation conditions	Proved in
NS, EN	**RAT**, **U**, **SYS**	[DL07a]
NS	**RAT**, **SYS**, **MON**	[NP10a]
PC, EN	**RAT**, **U**, **IND**	[DL07a, DH10a]
PC	**RAT**, **U**, **IND**, **MON**	[NP10a]

Figure 3.4: Combinations of agenda and aggregation conditions. If the agenda has the property on the left, then the property of the aggregation (middle column) is equivalent to dictatorship. The first row corresponds to Theorem 3.7.

In other words, all possible aggregation functions are dictatorships (or oligarchies).

Other theorems analogous to Theorem 3.7 (or Theorem 3.8) can then be obtained by varying the logical strength of the agenda and aggregation conditions considered, e.g., by strengthening EN with PC and weakening at the same time **SYS** to **IND**. Figure 3.4, which we adapted from [Lis12], recapitulates in a compact way some of the better-known impossibility results that have been established in the judgment aggregation literature.[12] Note that **RAT** is a constant assumption.

The third line of Figure 3.4 deserves special mention, stating that if the agenda is evenly negatable and path-connected then the conditions of collective rationality, unanimity and independence force the aggregation to be dictatorial.[13] This is a generalization of Arrow's theorem, whose agenda $\pm \{a \succ b, b \succ c, c \succ a\}$ (recall Examples 2.12 and 2.15) satisfies NS and PC. So, in the judgment aggregation setting, Arrow's theorem reads as follows:

Let $\mathcal{J} = \langle N, A \rangle$ be a judgment aggregation problem where $A = \pm \{a \succ b, b \succ c, c \succ a\}$. An aggregation function f satisfies **RAT**, **U** and **IND** if and only if it is a dictatorship.

One more impossibility result of special relevance will be studied later in Chapter 5.

3.4.2   INFINITE AGENDAS AND INFINITE VOTERS

In this subsection[14] we comment on what happens to the judgment aggregation problem when we relax the assumptions that the agenda and the set of voters are finite (Definition 2.1). For the first, the answer is simple. All results treated in this (and later) chapters carry over to the case of infinite agendas. Suffice it to notice that the finiteness assumption over A has never played a role in the proofs we presented.

---

[12]It is worth mentioning that for all rows except the first one in the table in Figure 3.4, the converse direction of the statement also holds. When that is the case, such impossibility results are also known as *agenda characterization theorems*.

[13]The very first generalization of Arrow's theorem along these lines—but requiring also the monotonicity condition—was proved in [Neh03].

[14]We are indebted to Umberto Grandi for insightful discussions on the contents of this subsection.

The finiteness assumption over N, on the other hand, plays a critical role. The pivotal step in the dictatorship- and oligarchy-based impossibility results handled in this chapter (Theorem 3.7 and Theorem 3.8), and later in Chapter 5 (Theorem 5.12), is enabled by Lemmas 3.6 and 3.10: on a finite domain, filters always contain a smallest set (the oligarchy), and ultrafilters always contain a singleton (the dictator). As already recognized by the very first papers applying the ultrafilter technique to obtain Arrow-like impossibility results in preference aggregation [Fis70, KS72, Han76], the two lemmas go through because of the finiteness assumption over the set of voters. Dropping that assumption—and thereby studying the problem of aggregation over infinite electorates—makes non-degenerate aggregation possible. Infinite electorates are not just a mathematical diversion, and apply whenever it is reasonable to assume the number of voters to be unbounded, like for instance in aggregation problems involving a group of individuals and all their future generations [Koo60].

The study of judgment aggregation on infinite electorates is an area of active ongoing research [HE12, Her12]. Here we give and prove two simple examples of possibility results for infinite voters.

Theorem 3.11   Non-oligarchic aggregation with infinite electorates.   *Let $\mathcal{J} = \langle N, A \rangle$ be an aggregation problem where $|N|$ is infinite. There exists an aggregation function f which is consistent, closed, satisfies* **U**, **SYS** *and does not satisfy* **O**.

Proof. We prove the claim by construction. Define the aggregation function f as follows:[15]

$$\varphi \in f(P) \quad \text{IFF} \quad P_\varphi \text{ is co-finite} \qquad (3.3)$$

for $P \in \mathbf{P}$ and $\varphi \in A$. It is clear that f satisfies **U** and **SYS**. Notice now that the set of winning coalitions $\mathcal{W}_f$ (Formula 3.1) defined by f is a proper filter: (i) $N \in \mathcal{W}_f$ and (ii) $\emptyset \notin \mathcal{W}_f$, since N is co-finite; (iii) $\mathcal{W}_f$ is upward closed, since each superset of a co-finite set is also co-finite; (iv) $\mathcal{W}_f$ is closed under finite intersections, since the intersection of two co-finite sets is also co-finite (the reader is invited to check this is the case). Since $\bigcap \mathcal{W} = \emptyset \notin \mathcal{W}_f$, by (ii), there exists no smallest O in $\mathcal{W}_f$ and hence f does not satisfy **O**. We need to show that, for any $P \in \mathbf{P}$, $f(P)$ is consistent and deductively closed. $\boxed{\text{Consistency}}$ Assume toward a contradiction that $f(P)$ is inconsistent (notice that such set is finite[16]). By (ii) and (iv) above we have that $\emptyset \neq \bigcap_{\varphi \in f(P)} P_\varphi \in \mathcal{W}$, from which we conclude that some voters hold inconsistent judgment sets, against Definition 2.2. $\boxed{\text{Closure}}$ The proof is similar to the one for consistency. $\square$

---

[15]We remind the reader that a co-finite set is an infinite set whose complement is finite. The set of co-finite sets of a given set N is commonly known as the *Fréchet filter* on N.

[16]If the agenda was infinite then, by the compactness of propositional logic (recall footnote 15 in Chapter 2) it would suffice to consider a finite inconsistent subset of $f(P)$.

The theorem is proven by constructing a specific aggregation function (Formula 3.3) working by the following principle: if there is an infinity of individuals accepting φ and only a finite number of them who reject φ, then φ should be collectively accepted. Intuitively, the function can be seen as a particular interpretation of how voting by majority could work in the presence of infinite voters.[17] Just like the propositionwise majority rule in presence of an electorate which is split in half, the above rule also yields incomplete collective sets in profiles where infinite individuals accept and reject some issue.[18]

Based on Theorem 3.11 one can obtain a similar possibility result for non-dictatorial aggregation:

Theorem 3.12 Non-dictatorial aggregation with infinite electorates.   *Let $\mathcal{J} = \langle N, A \rangle$ be an aggregation problem where $|N|$ is infinite. There exists an aggregation function f which satisfies* **RAT,** **U, SYS** *and does not satisfy* **D.**

Sketch of proof. By a result due to Tarski [Tar30], each filter can be extended to an ultrafilter. It follows that the filter $\mathcal{W}_f$ defined by the aggregation function f of Formula 3.3—the Fréchet filter—can be extended to an ultrafilter $\mathcal{W}'_f \supset \mathcal{W}_f$. Observe that $\mathcal{W}'_f$ is not principal, for otherwise we would have $\{i\} \in \mathcal{W}'_f$ (for some $i \in N$) and $N - \{i\} \in \mathcal{W}'_f$ and hence $\emptyset \in \mathcal{W}'_f$, and $\mathcal{W}'_f$ would not be an ultrafilter. We can now define:

$$\varphi \in f'(\varphi) \quad \text{IFF} \quad P_\varphi \in \mathcal{W}'_f \tag{3.4}$$

for $\varphi \in A$ and $P \in \mathbf{P}$. Clearly, f' satisfies **U** and **SYS**, and since $\mathcal{W}'_f$ is not principal it does not satisfy **D**. It remains to be shown that $f'(P)$ is a judgment set. $\boxed{\text{Consistency}}$ The consistency claim is proven like in the proof of Theorem 3.11. $\boxed{\text{Completeness}}$ Since $\mathcal{W}'_f$ is an ultrafilter, then for any $\varphi \in A$ we have that, for any $P \in \mathbf{P}$ either $P_\varphi \in \mathcal{W}'_f$ and hence $\varphi \in f'(P)$, or $P_{\neg\varphi} \in \mathcal{W}'_f$ and hence $\neg\varphi \in f'(P)$. $\qquad\qquad\square$

The two above theorems illustrate the possibility of democratic and rational aggregation when the number of individuals is infinite, independently of the logical structure of the agenda. In fact, neither of the two results assumes any agenda condition.

Infinite electorates can thus rightly be seen as a route, although rather limited, to the possibility of non-degenerate aggregation. In the next chapter we will review some of the other proposals that have been put forth in the literature to circumvent the constrictions charted by impossibility theorems. But before that we conclude this chapter with one last comment on the relationships between preference and judgment aggregation.

---

[17]Other, more subtle, interpretations are possible. We refer the reader in particular to [PS04] and [Fey04].

[18]For instance, consider the case when N is the set of natural numbers and all odd voters accept φ whereas all even voters accept $\neg\varphi$.

3.4.3   JUDGMENT AGGREGATION VS. PREFERENCE AGGREGATION

We have seen earlier in Section 1.2.2 how the Condorcet paradox can be rephrased as a judgment aggregation paradox. Since the early days of judgment aggregation, a natural question has been whether the theory could be shown to subsume preference aggregation. The question was answered positively in [DL07a], which showed how Arrow's theorem can be obtained as a corollary of a judgment aggregation theorem (see Section 3.4.1 above).

So Arrow's theorem is an instance of a more general judgment aggregation impossibility result. But can we go the other way around too? That is, can we view judgment aggregation impossibility results as instances of preference aggregation impossibilities? This question has been partially investigated in [Gro09, Gro10] and has been given a first positive answer. That work provides a number of results at the interface of judgment aggregation, preference aggregation and many-valued logics (see, for instance, [H01]) and is based on the following simple observation: i) preferences (strict $\succ$ and weak $\succeq$ ones) can be studied in terms of numerical ranking functions u, e.g., on the $[0, 1]$ interval [Deb54]; ii) numerical functions can ground logical semantics, like it happens in many-valued logic [H01] where, like in propositional logic, the semantic clause $u(x) \leq u(y)$ typically defines the satisfaction by u of the implication $x \to y$:

$$u \models x \to y \text{ iff } u(x) \leq u(y). \tag{3.5}$$

Intuitively, implication $x \to y$ is true (or accepted, or satisfied) iff the rank of x is at most as high as the rank of y. Preference aggregation (on possibly weak preferences) can then be studied as an instance of judgment aggregation on many-valued logics. In turn, judgment aggregation can be studied as an instance of a type of preference aggregation defined on dichotomous preferences, thus enriching the picture of the logical relationships between preference aggregation and judgment aggregation.

CHAPTER 4

Coping with Impossibility

In the lecture delivered when he received the Alfred Nobel Memorial Prize in Economic Sciences in 1998, Amartya Sen touched upon the proximity of possibility and impossibility results in social choice theory. When we introduce a set of axioms, there may exist several aggregation procedures that satisfy them. By introducing further axioms, we can reduce the number of possible procedures until we are eventually left with only one possibility.

> We have to go on cutting down alternative possibilities moving—implicitly— *toward* an impossibility, but then stop just before all possibilities are eliminated [...]. Thus, it should be clear that a full axiomatic determination of a particular method of making social choice must inescapably lie next door to an impossibility. [...] It is, therefore, to be expected that constructive paths in social choice theory, derived from axiomatic reasoning, would tend to be paved on one side by impossibility results (opposite to the side of multiple possibilities). [...] The real issue is not, therefore, the ubiquity of impossibility [...], but the reach and reasonableness of the axioms to be used. [Sen99, p. 354]

Naturally, such proximity of possibility and impossibility results holds also for judgment aggregation. Impossibility results such as the ones studied in Chapter 3 are usually seen as negative results. However, they have a more positive side in that they indicate which conditions may be relaxed in a quest for possibility results.

Chapter 3 has shown that at the heart of impossibility results lie two types of axioms: agenda conditions and aggregation conditions, the latter consisting of output and mapping conditions. So possibility results may be obtained once any of these types of conditions are relaxed. Among these, however, agenda conditions appear to be ineludible. Even the weakest of such conditions, non-simplicity, suffices for yielding impossibility results.[1] Relaxing agenda conditions would therefore amount to restricting judgment aggregation to somewhat trivial decision problems. So, following the structure of [Lis12], escape routes have to be found in: relaxing the output conditions; relaxing the mapping conditions, and more specifically **IND**; or, in addition, relaxing the universal domain condition built in the definition of aggregation function (recall Remark 2.5).

Chapter outline: Section 4.1 presents the results obtained when we restrict the domain of the aggregation function, while Section 4.2 reviews what happens when we relax collective rationality.

---

[1]Nehring and Puppe, for example, showed that, if the agenda is non-simple, every aggregation function satisfying **RAT**, **SYS** and **MON** is a dictatorship [NP10a] (this is one of the theorems in Figure 3.4).

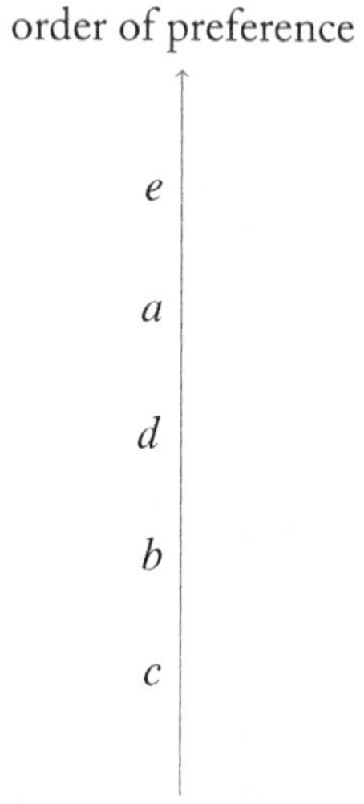

order of preference

e

a

d

b

c

Figure 4.1: A preference over the set of candidates $\{a, b, c, d, e\}$.

Finally, in Section 4.3 we present the third investigated escape route in judgment aggregation, which consists of dropping the independence condition. The closing section briefly reports on work geared toward the relaxation of yet other types of constraints, and on a specific application of non-independent aggregation within artificial intelligence. The chapter builds on material and results presented in [KPP99, Lis02, Gär06, DL07b, DM10, Pig06, CP11, CPP11].

4.1 RELAXING UNIVERSAL DOMAIN

4.1.1 UNIDIMENSIONAL ALIGNMENT

Individual preferences can be represented in several ways. One way is to align the alternatives in decreasing order over a line, like in Figure 4.1, which represents the following preference ordering: $e \succ a \succ d \succ b \succ c$.

Another representation is to place the alternatives on a horizontal axis and to mark the relative order of preference on the vertical axis. The same individual representation $e \succ a \succ d \succ b \succ c$ is shown as the solid line in Figure 4.2. As observed by Duncan Black [Bla48], when dealing with the preferences of a single individual, it will always be possible to choose an ordering of the alternatives on the horizontal axis such that the preference curve has a simple shape. However, when we wish to represent the preferences of several individuals on the same two-dimensional diagram, it may not always be possible to obtain simple shapes. In 1948 Black discovered that the curves representing the preferences of a group's members can tell us something significant.

He observed that in many practical group decision problems, alternatives can naturally be aligned along a left/right dimension. For example, when alternatives are political issues individuals

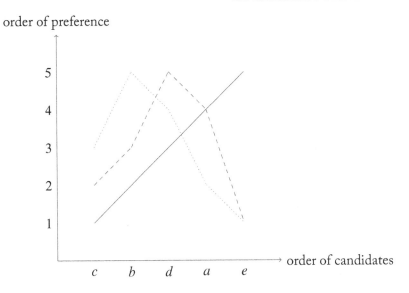

Figure 4.2: Single-peaked preferences over the set of candidates $\{a, b, c, d, e\}$.

often agree on an order from 'most conservative' to 'most liberal'. Similarly, when alternatives are numerical quantities (like the price of a product or the rate of a particular tax), it is easy to find a left/right ordering of the same alternatives. In all those cases, each individual tends to identify her optimum (the alternative that she prefers the most) and to favor less any alternative that lies the farther from her optimum. When represented on a two-dimensional diagram, the individual preferences have the shape of a *single-peaked curve*. These are individual preferences where there is a peak, which represents the most preferred alternative. On either side of the peak lie the less preferred alternatives (unless the peak is at the extreme left or right). The alternatives are ordered in such a way that their desirability declines the farther they are from the peak.

Preferences in Figure 4.2 are single-peaked. The five candidates (a, b, c, d and e) are ordered on the horizontal axis in such a way that the preferences of the three voters have a peak. Without referring to a spatial representation, we can say that preferences are single-peaked if, for any triple, there is an alternative that all individuals agree is not the worst. This is not the case, for instance, for the voters of the Condorcet paradox we encountered in Section 1.2.2 whose preferences are not single-peaked.

The so-called *median voter theorem* [Bla48] shows that, when single-peakedness is satisfied, the pairwise majority rule selects the alternative that received the highest number of votes, i.e., a Condorcet paradox cannot arise. This theorem says that if the number of voters is odd, all voters' preferences are single-peaked and there is a finite number of alternatives, then the peak of the median voter is the *Condorcet winner*. The median voter in Figure 4.2 is the voter represented by

	Voter 3	Voter 2	Voter 5	Voter 4	Voter 1
p	0	0	0	1	1
q	1	1	0	0	0
r	1	0	0	0	0

Figure 4.3: An example of a unidimensionally aligned profile in judgment aggregation [Lis02].

the dashed line. Thus, alternative d is elected. From Section 1.1.1 we recall that in an election the *Condorcet winner* is the candidate that receives the highest number of votes in all pairwise comparisons. The Condorcet paradox shows that a Condorcet winner does not always exist. Black's theorem is important because it shows that, if we can restrict the domain of possible individual orderings, then we can avoid the Condorcet paradox.[2]

A generalization of Black's result to *any* number of alternatives was attained by Arrow [Arr63, Chapter VII]. We note that single-peakedness is a *sufficient* condition on the preference profile to obtain a possibility result under majority voting. Necessary and sufficient conditions have been later studied by Sen and Pattanaik [SP69].

Researchers have explored whether it was possible to transpose the idea of single-peaked preferences to judgment aggregation. As seen in Definition 2.4, a judgment aggregation function takes as input a profile of consistent and complete subsets of the agenda. This means that the domain of the aggregation function is the set of all possible judgment sets (the so-called *universal domain* condition).[3] Inspired by Black's theorem, List [Lis02, Lis05a] introduced a similar condition to single-peakedness for judgment aggregation, called *unidimensional alignment*:

Unidimensional alignment. A profile P is *unidimensionally aligned* if there exists a strict linear order $>$ such that, $\forall \varphi \in A$: it is *either* the case that $\forall i, j \in N$ if $i \in P_\varphi$ and $j \in P_{\neg\varphi}$ then $i > j$, *or* it is the case that $\forall i, j \in N$ if $i \in P_\varphi$ and $j \in P_{\neg\varphi}$ then $j > i$.

I.e., the voters can be ordered from left to right in such a way that, for each formula in the agenda, the voters accepting that formula are either all to the left or all to the right of the voters rejecting it.

The profile seen in Figure 1.4 to illustrate the doctrinal paradox is not unidimensionally aligned. An example of a unidimensionally aligned profile is given in Figure 4.3.

List showed that, under such a domain restriction, propositionwise majority voting is the only aggregation procedure that guarantees complete and consistent collective judgment sets and that satisfies **SYS** and **AN**.

The reason why unidimensional alignment is sufficient for reaching consistent collective sets is that individuals are ordered in such a way that those accepting a formula are opposite

---

[2]Moulin [Mou80] showed that the restriction to single-peaked profiles can also ensure non manipulable aggregation functions, a topic to which we turn in Chapter 5.

[3]Recall Remark 2.5.

those rejecting that same formula. Thus, if the number of individuals is odd, the majority rule must coincide with the median voter's judgment set (Voter 5 in Table 4.3). Since we assume that individuals are logically consistent, so must be the collective set. If there is an even number of individuals, the majority will be the intersection of the judgment sets of the two median voters (which will still be a consistent set). What may happen is that the two median voters do not agree on an issue φ, that is, one may accept φ while the other $\neg\varphi$. In that case, the collective set would not be complete.

4.1.2 VALUE-RESTRICTION

The exploration of domain restriction conditions in judgment aggregation that guarantee possibility results continued in [DL10b]. Here, Dietrich and List introduced other sufficient conditions for majority consistency. In particular, they generalized another well-known condition in the theory of preference aggregation: the value-restricted preferences. Introduced by Sen [Sen66], this condition is more general than Black's single-peakedness.

Considering "concerned individuals," that is, individuals who are not indifferent between all the alternatives, Sen defines the value of an alternative in a triple, for a given preference, as being "best," "worst," or "medium." The assumption of value-restricted preferences is then expressed as follows:

A set of individual preferences over a triple of alternatives such that there exist one alternative and one value with the characteristic that the alternative never has that value in any individual's preference ordering, is called a value-restricted preference pattern over that triple for those individuals. [Sen66, p. 492]

This means that, given a triple of alternatives, there is some alternative over which all concerned individuals agree it is not best, or agree that it is not worst, or agree that it is not medium [Gae06, p. 44]. Clearly, value restriction is violated in the Condorcet paradox (recall Example 1.1).

The translation of the above condition in the context of judgment aggregation led Dietrich and List to formulate the *value-restricted* condition below:

Value-restriction. A profile P is value-restricted if for every minimal inconsistent set $X \subseteq A$ there exists a two-element subset $Y \subseteq X$ that is not accepted by *any* individual $i \in N$.

Value-restriction can be seen as an agreement among individuals that, for every minimal inconsistent subset of the agenda, there are two propositions that nobody in the group supports together. Value-restriction is sufficient to avoid that an inconsistent majority judgment set is selected as the group outcome.

Example 4.1   To illustrate this condition, let us consider the agenda $\pm\{p, q, p \vee q\}$. A minimal inconsistent subset is $\{p \vee q, \neg p, \neg q\}$. The value-restriction condition says that there exists a conjunction of two propositions of that subset (e.g., $(p \vee q) \wedge \neg p$, $(p \vee q) \wedge \neg q$, or $\neg p \wedge \neg q$) that is not satisfied by any judgment set in the profile. Thus, for example, the following profile is value-restricted (it does not contain $\{p \vee q, \neg q\}$) and does not lead to paradoxical outcomes.

	p	q	$p \vee q$
V_1	0	0	0
V_2	0	1	1
V_3	0	1	1
f_{maj}	0	1	1

Dietrich and List also introduce a *necessary and sufficient domain-restriction condition for majority consistency*:

Majority consistency. A profile P is majority consistent if every minimal inconsistent set $X \subseteq A$ contains a proposition that is not accepted by a majority.

Domain-restriction conditions can represent plausible escape-routes to the impossibility results in some decision-making contexts. As observed by List [Lis12], different groups display different levels of pluralism. If there is empirical evidence showing that the conditions above are met in a specific group confronted with a particular decision problem, then individual judgments can be safely aggregated into a collective judgment set. For a critical discussion of Sen's value-restriction condition and other domain restrictions, see [RGMT06].

4.2   RELAXING THE OUTPUT CONDITIONS

A crucial requirement on aggregation functions is that their output be rational, that is, a consistent and complete set of formulae. Consistency is usually seen as an indispensable requirement.[4] The other requirement of collective rationality, namely completeness, may be given up in those situations in which a decision on all the agenda's issues is not strictly required. Two main ways to relax completeness have been explored in the literature: the first allows individuals to abstain, the second resorts to quota rules.

4.2.1   ABSTENTION

Relaxing completeness, at both the individual and collective level, has been explored as a way to avoid impossibility results and to model more realistic decision procedures. Gärdenfors [Gär06] criticized the completeness requirement as being too strong and unrealistic. But what happens when voters are allowed to abstain from expressing judgments on some propositions in the agenda? Gärdenfors proved that, on specific agendas, if the judgment sets need not to be complete

---

[4]For a recent proposal to drop consistency and obtain a positive result, see Section 4.4.2.

but are deductively closed (recall Remark 2.3) and consistent, then every aggregation function that is **IND** and **U** must be *weakly oligarchic*,[5] that is:

Weakly oligarchic iff $\exists O \subseteq N$ s.t. $O \neq \emptyset$ and $\forall(J_1, \ldots, J_n) \in \mathbf{P} : f(J_1, \ldots, J_n) \supseteq \bigcap_{i \in O} J_i$.

I.e., an aggregation function is *weakly oligarchic* (or a *weak oligarchy*) if there exists a non-empty smallest subset of voters O such that the collective set always contains all formulae on which all individuals in O agree.[6]

Clearly, when O contains only one member, oligarchy reduces to dictatorship. At the other end of the spectrum, when $O = N$, we have the unanimity rule (discussed earlier in Subsection 2.1.4), which can thus be seen as the oligarchy of the whole set of individuals N. It is worthwhile noticing that, when $O = N$, the decision procedure is anonymous but only unanimous issues are upheld by the group.[7] Gärdenfors seems to favor those kinds of rules:

> [S]ince an oligarchy will only be fully democratic in the limiting case when it consists of all members of the voting community, the theorem [...] points to unanimous voting functions as the only acceptable ones. [Gär06, p. 189]

Gärdenfors's framework requires the agenda to have a very rich logical structure (with an infinite number of issues). Later, Dokow and Holzman [DH10b] considered abstention with finite agendas. Simpler agendas were also assumed by Dietrich and List [DL08], as in the impossibility of non-oligarchic aggregation which we proved in Chapter 3 (Theorem 3.8). All these results show that Gärdenfors's insights can be obtained with weaker conditions on the aggregation functions, and that in fact these conditions lead to oligarchies, and not merely weak oligarchies. Unlike Gärdenfors's, Dietrich and List's conclusion is rather pessimistic: if the goal is to avoid impossibility results, dropping completeness does not lead us very far as dictatorial rules are simply replaced by oligarchic ones.

4.2.2 QUOTA RULES

In this section, we come back to the threshold-based rules we have introduced in Chapter 2 (Subsection 2.1.4) as *quota rules* and explore the sort of avenues they offer in mitigating the impact of impossibility results.

When quota rules are used, a proposition is in the collective set if and only if that proposition is accepted by a number of individuals greater than a prefixed threshold. The appeal of quota rules comes from the intuition that different problems may require different social support in order to be declared collective decisions. For example, a decision that has a high impact on a

---

[5]Notice that this result, unlike Theorem 3.8, assumes that also individual judgment sets and not only the collective ones may be incomplete. Nonetheless the theorem can still be proven through the filter technique used in Chapter 4. The reader is invited to consult [Gär06] for details.

[6]It is worthwhile to compare this definition with the definition of oligarchic function given in Chapter 2.

[7]Notice that the unanimity rule guarantees deductive closure at the expense of significant incompleteness.

group may require being supported by 2/3 of the individuals rather than by a simple majority. Furthermore, in a given agenda, one issue may be more important than another and so different propositions may have different thresholds. Majority voting, we have seen, is a special kind of quota rule, with the same majority threshold for each proposition. Clearly, quota rules do not guarantee complete collective sets.

Dietrich and List [DL07b] explored quota rules in the context of judgment aggregation. They considered four rationality conditions on the individual and collective judgments. Besides completeness and consistency, they considered deductive closure and weak consistency, which is the property demanding that a proposition and its negation cannot be accepted at the same time. They showed that a given quota rule satisfies a rationality condition if a certain inequality concerning the thresholds is verified. We have already seen simple examples of such inequalities (Formulae 2.4 and 2.5) when discussing quota rules in Chapter 2. Whether such inequalities are satisfied depends on the logical structure of the agenda (in particular, on the size of the minimal inconsistent subsets of the agenda). For rich agendas, these turn out to be rather demanding conditions, and even more so if we require that two conditions are met at the same time (for example, consistency and deductive closure).

Nevertheless, it is worth mentioning that, if we drop completeness and weaken collective rationality to consistency alone, supermajority rules produce consistent collective judgments when the supermajority threshold q for every proposition is greater than $n - (\frac{n}{k})$, where n is the number of individuals and k is the size of the largest minimally inconsistent subset of the agenda. As k increases, the threshold to ensure consistency approaches n (thus, requiring unanimity). Furthermore, for groups with at least three voters, the propositionwise majority rule is consistent and deductively closed only when $k \leq 2$, that is, if the agenda is simple (cf. Theorem 3.2).

4.3   RELAXING INDEPENDENCE

All impossibility results we touched upon feature **IND** as a central condition, or its strengthening **SYS**.[8] **IND** rephrases in the context of judgment aggregation the *independence of the irrelevant alternatives* condition of Arrow's theorem. Independence of irrelevant alternatives warrants that the group ranking over any pair of alternatives depends solely on the individual rankings over the same pair of alternatives. The intuition is that the social ranking over, for example, x and y should be determined exclusively by how the individuals rank x compared to y and not by the ranking of other (irrelevant) alternatives like, for instance, z. The analogous requirement in judgment aggregation ensures that the collective judgment on each proposition depends exclusively on the individual judgments on that proposition.

As we will see in Section 5.1.1 (and more specifically in Theorem 5.5), **IND** is a key condition to ensure that an aggregation function is non-manipulable [Die06, DL07c], i.e., robust

---

[8]Recall that **SYS** is the conjunction of **IND** and **NEU** (Section 2.3.2).

against strategic voting. This makes **IND** an instrumentally attractive condition, like the independence of the irrelevant alternatives condition in preference aggregation.[9] However, **IND** has also been severely criticized in the literature (see, for example, [Cha02, Mon08]). Several authors deem **IND** incompatible with a framework whose aim is precisely the one of aggregating logically *interrelated* propositions. Mongin, for example, writes:

> [...] the condition remains open to a charge of irrationality. One would expect society to pay attention not only to the individuals' judgments on φ, but also to their *reasons* for accepting or rejecting this formula, and these reasons may be represented by other formulas than φ in the individual sets. [Mon08, p. 106]

These criticisms make **IND** a plausible candidate for a condition to be relaxed in order to achieve possibility results. In this section we will consider three main options to relax **IND**: the premise-based approach, the sequential priority approach, and the distance-based approach.

4.3.1   THE PREMISE-BASED APPROACH

When Kornhauser and Sager discovered the doctrinal paradox, they observed that there are two plausible ways in which a court can overcome the *impasse* and reach a decision under majority rule: either by the *issue-based* method [KS93], *a.k.a.* the premise-based procedure, or by the *case-based* method, *a.k.a.* the conclusion-based procedure.[10] If at the beginning the premise-based approach was viewed as one possible workaround the doctrinal paradox, with the appearance of the first impossibility theorem [LP02] it provided also an argument for relaxing **IND** and thus escaping some of the impossibility results.

In the premise-based procedure the agenda is assumed to be partitioned into two disjoint subsets: *premises* and *conclusions*, and the premises are usually assumed to be logically independent (otherwise a discursive paradox could arise over them).[11] Individuals express their judgments on the premises only. The collective set contains the propositionwise aggregation (e.g., through the majority rule) of the individual judgments on the premises. From the collective outcome on the premises, the collective conclusions are derived using the logical relationships between the agenda issues. This means that some propositions (the premises) are prioritized over others (the conclusions), making this procedure a special case of the sequential priority approach that we will

---

[9]Regarding this issue it might be worth mentioning, in passing, an interesting position that has been expressed by Dowding and van Hees [DvH07]. Their claim is that strategic voting is not necessarily a vice. One of their claims is that, in order to manipulate, individuals need to understand the voting system, which constitutes a virtue as it provides incentives for the comprehension of the democratic procedures.

[10]We have introduced these procedures in Chapter 1 and formally defined them in Section 2.1.4.

[11]In reality, there are different definitions of premise-based procedures in the literature, depending on which of the following features (and their combinations) are assumed: some take premises to be logically independent [LP02, Neh05, Die06, Mon08, NP10b], some assume that premises fully determine the conclusions (thus guaranteeing complete judgment sets as outcomes) [Neh05, NP06, Mon08], and some that there is only one conclusion [Neh05, NP10b]. If the premises are not assumed to be logically independent, then the majority rule on the premises is ensured to return a consistent judgment set if the sub-agenda constituting the premises is simple (Theorem 3.2).

explore in the next section. The open question, however, remains of how to partition an agenda into premises and conclusions in a principled manner.

Premise-based vs. conclusion-based

In Section 1.2.1 we have also considered the conclusion-based procedure and we have noticed that it may give an opposite result than the premise-based method. This incompatibility, we have seen, was at the heart of the legal theory debate on the doctrinal paradox and the discursive dilemma. So the question remains of how to choose between the two approaches. One possible answer has been given by Bovens and Rabinowicz [BR06] and by List [Lis05b]. The idea is to evaluate and compare the two aggregation procedures in their *truth-tracking* reliability. It is assumed that a group judgement is factually right or wrong and, therefore, the question is how reliable the two approaches are at selecting the right judgment set.

If the individuals are better than randomizers at judging the truth or falsity of a proposition (in other words, if the probability of each agent at getting the right judgment on a proposition is greater than 0.5), and if they form their opinions independently, then the probability that majority voting yields the right collective judgement on that proposition increases with the increasing size of the group. As we know from Chapter 1, this was one of Condorcet's findings—the *Condorcet Jury Theorem*—which links the competence of the agents to the reliability of majority voting. It also motivates the use of majority-based decision making in the judgement aggregation problem. The interesting result of [BR06, Lis05b] is that the premise-based procedure is a better truth-tracking approach than the conclusion-based procedure.

Drawbacks of the premise-based procedure

Despite all these good news, the premise-based procedure can lead to unwelcome results. Because the collective judgment on the conclusion is derived from the individual judgments on the premises, it can happen that the premise-based procedure violates a unanimous vote on the conclusion. Nehring [Neh05] presents a variation of the discursive dilemma, which he calls the *Paretian dilemma*. In his example, a three-judges court has to decide whether a defendant has to pay damages to the plaintiff:

> Legal doctrine requires that damages are due if and only if the following three premises are established: 1) the defendant had a duty to take care, 2) the defendant behaved negligently, 3) his negligence caused damage to the plaintiff. [Neh05, p. 1]

Suppose that the judges vote as in Table 4.4. The Paretian dilemma is disturbing because, if the judges follow the premise-based procedure, they condemn the defendant to pay damages contradicting the *unanimous* belief of the court that the defendant is *not* liable. The trouble is that all anonymous or non-dictatorial aggregation functions are prone to the Paretian dilemma [Neh05]. How negative is this result? Nehring argues that when the reasons are epistemically independent:

	p	q	r	$x = (p \land q \land r)$
Judge 1	1	1	0	0
Judge 2	0	1	1	0
Judge 3	1	0	1	0
Majority	1	1	1	0

Figure 4.4: Paretian dilemma. Premises: p = duty, q = negligence, r = causation. Conclusion: $x = (p \land q \land r)$ = damages.

all relevant information about the outcome decision is contained in the agents' premise judgments. [...] Indeed, under epistemic independence of premises it is easy to understand how a group aggregation rule can *rightly* override a unanimous outcome judgment. [Neh05, p. 36].

Furthermore, the normative force of the Pareto criterion (which corresponds to our unanimity condition **U**) depends on the type of social decision. The Pareto criterion should be ensured when the individuals have a *shared self-interest* in the final outcome, whereas it can be relaxed when they *share responsibility* for the decision. Judicial decisions are clear instances of shared responsibility situations, while other group decisions may be self-interest driven. Nehring's analysis concludes that the Pareto criterion and reason-based group decisions are two principles that may come into conflict.

A general premise-based procedure that neither assumes the premises to be logically independent, nor that they fully determine the conclusions has been later defined by Dietrich and Mongin [DM10]. They state necessary and sufficient conditions for any aggregation rule to be dictatorial on the premises (or on the whole agenda) when we restrict **IND** to the premises only and impose **U** on the whole agenda.[12] Whether the rule degenerates into dictatorship (or oligarchy) depends on the logical connections within the premises and between premises and conclusions.

4.3.2   THE SEQUENTIAL PRIORITY APPROACH

Another possibility to relax **IND** is the sequential priority approach. Sequential procedures [Lis04, DL07b] work in this way: the elements of the agenda are considered sequentially, following a fixed linear order over the agenda (corresponding, for instance, to temporal precedence or to priority), and earlier decisions constrain later ones. Thus, individuals vote on each proposition φ in the agenda, one by one, following the fixed order. If the collective judgment on φ is consistent with the collective judgments obtained on the previous issues of the agenda, the collective judgment on φ becomes the group position on φ. However, in case the collective position on φ conflicts with the group judgments on the propositions aggregated earlier, the collective judgment

---

[12]An oligarchic variant is obtained when the set of possible collective outcomes is taken to be the set of all judgment sets that are consistent and deductively closed, instead of consistent and complete.

	p	$p \leftrightarrow q$	q
President 1	1	1	1
President 2	0	0	1
President 3	0	1	0
Majority	0	1	1

Figure 4.5: An example of sequential majority rule.

on φ will be derived from the earlier group judgments. Clearly, the premise-based procedure is then a special instance of sequential priority procedures.

Collective consistency is guaranteed by definition. Of course, in the general case, the result depends on the choice of the order. This property is known as *path-dependency* of the rule. In order to illustrate how a sequential priority rule works and the problem of path-dependence, we recall here the example used in [DL07b].

Example 4.2 Sequential priority rules [DL07b]   Suppose that the presidents of three governments have to decide on the following propositions:

p: Country X has weapons of mass destruction.
q: Action Y should be taken against country X.
$p \leftrightarrow q$: Action Y should be taken against country X if and only if country X has weapons of mass destruction.

Suppose furthermore that the individual judgments on the issues in the agenda are as in Figure 4.5 and that simple propositionwise majority is used.

We can now consider two different sequential paths. In the first, the items of the agenda are aggregated according to the following order: $p, p \leftrightarrow q, q$. In the second path, agents are asked to vote in the following order: $q, p \leftrightarrow q, p$. We obtain two different collective judgments: $\{\neg p, p \leftrightarrow q, \neg q\}$ when the first path is followed and $\{p, p \leftrightarrow q, q\}$ when the second path is followed. In both cases, the three presidents agree that action Y should be taken against country X if and only if country X has weapons of mass destruction. However, while they will take action against country X if the first path is followed, they will take no action against country X if the second path is used.

Path-dependence is tightly linked to the manipulability of both the agenda and the vote in an aggregation problem, topics we will address in detail in the next chapter. The agenda-setter can manipulate the social outcome by fixing a specific order in which the items in the agenda are considered, and individuals may also have an incentive to misrepresent their own true judgments in order to force a collective outcome they favor.

4.3.3   THE DISTANCE-BASED RULES

The third approach that relaxes **IND** and that we consider here is the distance-based approach. Distance-based judgment aggregation rules [Pig06] have been originally derived from distance-based merging operators for belief bases introduced in computer science [KLM04, KG06, KPP99]. Unlike the premise-based procedure and the sequential priority approach, distance-based rules take as input a profile of judgment sets on the entire agenda.

Intuitions and formal definition

Distance-based rules assume a predefined distance between judgment sets and between judgment sets and profiles, and choose as collective outcomes the consistent and complete sets which are closest (for some notion of closeness) to the individual judgment sets. These outcomes are not necessarily unique and the rule is therefore irresolute (cf. Remark 2.5), unless a tie-breaking rule is applied to the output. Already from this informal presentation it is clear that distance-based rules do not satisfy independence. The group judgment on a proposition is not solely determined by the individuals' judgments on that proposition, but by considerations involving potentially all other propositions in the agenda.

Let us move to some formal definitions. Let $d : \mathbf{J} \times \mathbf{J} \longrightarrow \mathbb{R}^+$ be a distance function between any two judgment sets $J_i, J_j \subseteq X$.[13] Well-known are the *drastic* (or *Chebyshev*) *distance*, which assigns distance 0 if two judgment sets are the same and 1 otherwise, and the *Hamming* (or *Manhattan*) *distance*, which counts the number of propositions on which two judgment sets disagree. For example, if $J_i = \{p, \neg q, r\}$ and $J_j = \{\neg p, \neg q, r\}$, the Hamming distance d_H between the two judgment sets is 1 as they differ only on the evaluation of proposition p, that is, $d_H(J_i, J_j) = 1$. In the following we use the Hamming distance because of its intuitiveness and wide applicability. However, it should be stressed that this is only one among many possible distance functions that one may want to use [KPP99, KPP02b].

So, function d assigns a distance to each judgment set of a given profile P and any judgment set that can be selected to be the collective judgment set. Once all these distances are obtained, we need to calculate the distance between the profile and each possible collective judgment set. This is done with the help of a function $f : \mathbb{R}^{+^n} \longrightarrow \mathbb{R}^+$, which, given a profile of individual distances from the collective set, outputs a global distance. A simple and common example is $\sum_{i \in N} d(J_i, J)$, which obtains the global distance by summing up the individual ones.

The intuition behind such a distance-based rule consists in selecting those collective judgment sets that minimize the global distance from the judgment sets occurring in the profile. Formally, the distance-based rule $F^{d,f}$, where d is the Hamming distance and f is the sum,

---

[13]$\mathbb{R}^+$ denotes the non-negative reals. We recall that d is a distance function if and only if for all $J_i, J_j \subseteq X$ we have that: (i) $d(J_i, J_j) = d(J_j, J_i)$ and (ii) $d(J_i, J_j) = 0$ if and only if $J_i = J_j$. Technically, we slightly abuse terminology here, since d is only a so-called *pseudo-distance* as the triangular inequality ($\forall J_i, J_j, J_k \subseteq X, d(J_i, J_j) \leq d(J_i, J_k) + d(J_k, J_j)$) is not required to hold.

selects the judgment set that minimizes the sum of the Hamming distances (also called *minisum outcome* [BKS07]), and is defined as follows for $P \in \mathbf{P}$.

Minisum outcome rule:

$$\varphi \in F^{d_H, \Sigma}(P) \quad \text{IFF} \quad F^{d_H, \Sigma}(P) = \operatorname*{argmin}_{J \in \mathbf{J}} \sum_{i \in N} d_H(J_i, J) \tag{4.1}$$

The best way to illustrate how this particular distance-based rule works is with the help of an example.

Example 4.3 Distance-based aggregation Let us consider the doctrinal paradox. The three judgment sets corresponding to the three judges are:

$$
\begin{aligned}
J_1 &= \{p, q, r\} \\
J_2 &= \{p, \neg q, \neg r\} \\
J_3 &= \{\neg p, q, \neg r\}
\end{aligned}
$$

The table below shows the result of the distance-based aggregation rule defined in Formula 4.1. The first column lists all the consistent judgment sets. The numbers in the columns of $d(., J_1)$, $d(., J_2)$ and $d(., J_3)$ are the Hamming distances of each J_i from the correspondent candidate for collective judgment set. Finally, in the last column is the sum of the distances over all the individual judgment sets in the profile.

	$d_H(., J_1)$	$d_H(., J_2)$	$d_H(., J_3)$	$\sum(d_H(., P))$
$\{p, q, r\}$	0	2	2	4
$\{p, \neg q, \neg r\}$	2	0	2	4
$\{\neg p, q, \neg r\}$	2	2	0	4
$\{\neg p, \neg q, \neg r\}$	3	1	1	5

So, in this example, the consistent judgment sets that are closest to the profile P correspond exactly to the individual judgment sets in P (they are at distance 4 rather than 5). Thus, by restraining the choice of the outcome candidates to the consistent judgment sets, we avoid the paradox at the cost of adding irresoluteness to the aggregation.[14]

Belief merging and judgment aggregation

The aggregation of multiple databases, of potentially conflicting information from different sensors, as well as the combination of multiple knowledge bases in expert systems, are problems that have long occupied researchers in computer science [BKMS92, CGMH+94, ERS99]. In particular, the aggregation of independent and equally reliable sources of information expressed in logical

---

[14]Distance-based procedures have also been studied as truth-tracking procedures [HPS10, Wil09, HS11]. For a distance-based approach in a fuzzy framework, see [BK12].

form has been investigated in belief merging, from which distance-based judgment aggregation rules have been derived [KLM04, KG06, KPP99].

Different operators have been introduced and studied: for instance, combination operators that take the disjunction of the maximal consistent subsets [BKM91], arbitration operators that equally distribute the individual distances from the outcome candidates [Rev93, LS95, KPP02b], majority merging operators [LM99], and operators to merge prioritized knowledge bases [BDL+98].

New postulates for merging operators as well as the distinction between arbitration and majority operators were introduced in [KPP98]. The framework has been then extended [KPP99, KPP02a] to include merging under integrity constraints, that is, a set of exogenously imposed conditions that have to be satisfied by the merged base. Even if integrity constraints do not appear explicitly in Formula 4.1, it is thanks to them that the choice of collective judgment is restrained to those that are consistent (and complete).

Despite the transposition of aggregation methods from belief merging to judgment aggregation, there are some differences between fusing different belief bases and merging judgment sets. The first one is the lack of an externally given agenda in belief merging. As mentioned, aim of belief merging is to consistently aggregate the information coming from different sources, where such sources may have different access to the information or have different competences. This implies that not all belief bases will contain information on exactly the same propositions.

Another difference is that in belief merging no assumption about the consistency of the individual belief bases is made. If a belief base violates an integrity constraint, that base will not appear among the possible collective outcomes. However, its input will not be totally disregarded as distances from the untenable inputs to the admissible outputs will be calculated in the usual way.

Finally, unlike judgment aggregation that investigates the combination of one type of attitudes, belief merging aims at a more general framework, able to encompass the problem of aggregating symbolic inputs, without specifying whether such items are beliefs, knowledge, desires, norms etc. In belief merging, it is the choice of the merging operator that needs to best suit the type of inputs.

4.4 FURTHER TOPICS

We conclude by pointing the reader to some further lines of research in the quest for possibility results, and to a graph-based variant of judgment aggregation currently being investigated within the field of artificial intelligence.

4.4.1 MORE DOMAIN RESTRICTIONS

Besides value-restriction, Dietrich and List introduced other domain restrictions that guarantee consistent majority outcomes. In [DL10b], they introduced conditions based on orders of the propositions in the agenda (like single-plateauedness and single-canyonedness), and a condition of unidimensional orderedness that, like unidimensional alignment, is based on orders of individuals.

Domain restrictions are a form of social consensus. It is obvious that in a group where all individuals submit the same judgments, no inconsistency can arise. So, clearly, we may encounter paradoxes when the individuals constituting a group do not hold similar positions. In Chapter 7 we will look at deliberation in judgment aggregation. One open question is how a debate among voters may impact the individual opinions. More precisely, it would be interesting to understand under which conditions a deliberation before the vote may help the individuals to revise their opinions so that the resulting profile is, for instance, unidimensionally aligned. Similar investigations had been carried out in preference aggregation, looking at the so-called preference restrictions (beside value restriction and single-peakedness, other conditions have been introduced and studied, like *limited agreement* and *extremal restriction*, cf. [Gae01]).

An interesting recent strand of research is *behavioral social choice*, which compares empirical elections data with results in voting theory, with surprising results:

> Behavioral social choice research can bring a new perspective to Arrow's theorem if it demonstrates that *actual* (voting) data are such that majority rule is overwhelmingly transitive. [RGMT06, p. 4]

Regarding the domain restrictions introduced in the literature, like Black's single peakedness and Sen's value-restriction we mentioned in Section 4.1, Regenwetter et al. find that empirical preference distributions systematically violate any domain restriction condition. The reason is that, at least for mass electorates, any preference that an individual is allowed to submit, will be submitted by some individuals (and possibly by a large number of people), no matter how 'strange' such preference may appear.

4.4.2 DROPPING CONSISTENCY

In relaxing output conditions, we only focused on completeness and disregarded consistency. The reason is that consistency of collective judgments is usually seen as an irrevocable requirement. Yet, a recent proposal to drop consistency has been put forward in [BCEF]. Their approach is inspired by the work of de Finetti [dF74] and Joyce [Joy09], and motivated by paradoxes like the lottery and preface paradox discussed in the formal epistemology literature. They introduce a new coherence requirement, such that a judgment set is coherent if it is not accuracy-dominated, i.e., if there is no other judgment set that contains strictly fewer inaccurate judgments. The authors in [BCEF] show that, although aggregation by majority does not always yield a consistent collective set, it does always yield a collective set which is coherent in the above sense.

4.4.3 OTHER DISTANCE-BASED RULES

It has been shown [EM05] that, for a preference agenda, the minisum outcome rule $F^{d_H,\Sigma}$ defined in Formula 4.1 is equivalent to the so-called *Kemeny rule*, a well-known preference aggregation rule [Kem59]. The fact [YL78] that the Kemeny rule is the only preference aggregation rule that is neutral, consistent and satisfies the Condorcet property,[15] might be adduced as a justification for the use of $F^{d_H,\Sigma}$ as a method for judgment aggregation [EP05].

However, as Duddy and Piggins observed [DP12], there is an inherent difficulty with the use of the Hamming distance in judgment aggregation problems. To use their example, the Hamming distance between $\{p, q, p \wedge q\}$ and $\{p, \neg q, \neg(p \wedge q)\}$ is 2. But $(p \wedge q)$ is a logical consequence of p and q, and $\neg(p \wedge q)$ is a logical consequence of p and $\neg q$. If those are the judgment sets of two individuals, their disagreement over $(p \wedge q)$ is a consequence of their disagreement over q, so the disagreement should be counted only once. Hence, the Hamming distance neglects that propositions are logically connected, leading in some cases to double counting. In order to overcome this problem, Duddy and Piggins propose an alternative distance. Given two judgment sets, their distance is defined as the smallest number of logically coherent changes needed to convert one judgment set into the other.

Four general methods for distance-based judgment aggregation, that do not commit to a specific distance metrics, have been developed by Miller and Osherson [MO09]. The first method is called *Prototype* and generalizes the minisum outcome rule by not confining it to the Hamming distance. *Endpoint* selects the closest (according to a given distance) judgment set to the possibly inconsistent collective judgment set. *Full* looks for the closest profiles that return a consistent propositionwise majority collective judgment set. Finally *Output*, the fourth method, also looks for the closest modified profile that yields a consistent collective outcome but, unlike *Full*, it allows the individual judgment sets in the modified profile to be inconsistent.

Example 4.3 showed the table of the distances between each individual judgment set and the 4 possible collective judgment sets—that is, the judgment sets that are consistent with the $(p \wedge q) \leftrightarrow r$ constraint in the doctrinal paradox. Yet, had we considered all 8 possible evaluations of the three atomic propositions, thus including sets of propositions that violate the above rule, we would have found out that there was only one judgment set at a minimal distance from the profile, and such judgment set was precisely the one selected by propositionwise majority voting, i.e., $\{p, q, \neg r\}$. Is this a coincidence? Not really. The equivalence between the outcome selected by propositionwise majority voting and the minisum outcome of Formula 4.1 has been pointed out in [BKS07]. One can of course define other distance minimization rules [MO09, KLM04, KPP99]. For example, another widely used distance-based aggregation rule is the *minimax*, which selects the collective judgment set that minimizes the maximal distance to the individual judgment sets [BKS07]. The intuition of minimax is to keep the disagreement with the least satisfied individual

---

[15]We recall that a preference aggregation rule satisfies the Condorcet property if, whenever an alternative x defeats another alternative y in pairwise majority voting, it can never be the case that y is ranked immediately above x in the social preference.

Figure 4.6: An argumentation framework.

Figure 4.7: The three possible extensions for the argumentation framework of Fig. 4.6.

at minimum, thus guaranteeing some degree of compromise. For a given profile, minisum and minimax may select two opposite collective outcomes [EK07].

4.4.4 JUDGMENT AGGREGATION AND ABSTRACT ARGUMENTATION

Argumentation theory has attracted intense interest in AI in the last two decades. We mention here some recent work on judgment aggregation and abstract argumentation to illustrate a different framework in which to investigate aggregation problems, and one in which the impossibility results of judgment aggregation can be avoided by relaxing the independence condition.

Judgment aggregation can be seen as an aggregation of individual evaluations of a given argumentation framework. An *argumentation framework* is defined by a set of arguments and a (binary) attack relation among them. Given an argumentation framework, argumentation theory identifies and characterizes the sets of arguments (*extensions*) that can reasonably survive the conflicts expressed in the argumentation framework, and therefore can collectively be accepted. In general, there are several possible extensions for a set of arguments and a defeat relation on them [Dun95].

For example, in the argumentation framework in Figure 4.6, we have that argument A attacks argument B, and that B attacks A. There are three possible extensions for this argumentation framework, namely those pictured in Figure 4.7. The black color means that the argument is rejected, white means that it is accepted and grey means that it is undecided, i.e., one does not take a position about it.

The general idea is that, given an argumentation framework, individuals may provide different evaluations regarding what should be accepted and rejected. The question is then how to obtain a collective evaluation from individual ones. The aggregation of individual evaluations of a given argumentation framework raises the same problems as the aggregation of individual judgments. Indeed, voting for each argument whether it is accepted, rejected or undecided[16] may result in an unacceptable extension, as the propositionwise majority voting may output an inconsistent collective judgment set.

---

[16]Notice that, technically, this can be viewed as a judgment aggregation problem in many-valued logics (cf. Section 2.4.2).

Among the first who applied abstract argumentation to judgment aggregation problems were Caminada and Pigozzi [CP11]. In an earlier work, Rahwan and Tohmé [RT10]—building on a general impossibility theorem from judgment aggregation—prove an impossibility result and provide some escape routes. Moreover, in [RL08] welfare properties of collective argument evaluation are explored.

The goal of [CP11] is not only to guarantee a consistent group outcome, but also that such outcome is *compatible* with the individual judgments. Caminada and Pigozzi stress that group inconsistency is not the only undesirable outcome. It may happen, for example, that the argument-by-argument majority rule selects as social outcome a consistent combination of reasons and conclusion that actually no member voted for (a remembrance of another voting paradox, the *multiple election paradox* [BKZ98]). [CP11] proposed and studied three operators that guarantee a group outcome which is 'compatible' with its members' judgments. Compatible refers to group decision-making in which any group member is able to defend the group decision without having to argue against his own private opinions. It is shown that, not only a collective consistent decision is guaranteed, but that this is also unique. The three operators do not satisfy (a suitably adapted notion of) independence. Furthermore, another property that is not satisfied is the preservation of a unanimously supported outcome, giving rise to a situation like the Paretian dilemma mentioned in Section 4.2. In a follow-up paper [CPP11], Pareto optimality and manipulability aspects of these operators have been investigated.

CHAPTER 5

Manipulability

Aggregation functions work by taking as input one judgment set for each single individual. And it is up to the individual which judgment set to submit to the function. So the natural question arises: would it be possible for an individual to force the collective acceptance of a specific issue, by concealing her true judgment set and submitting an aptly modified one? This is, in a nutshell, the issue of the *manipulability* of judgment aggregation and it is the topic of the present chapter. We will study (non-)manipulability as a property of aggregation functions and explore the effects that such a property has on the process of aggregation.

Chapter outline: Section 5.1 introduces the issue of manipulation in its two main forms of agenda and vote manipulation, the latter being the main topic of the present chapter. It then provides a formal definition and characterization of the vote manipulability of a judgment aggregation function. Section 5.2 presents an impossibility result connecting non-manipulability to dictatorship in the spirit of Theorem 3.7 of Chapter 3. A detailed proof of the result is given highlighting another application of the axiomatic method and the ultrafilter technique in the context of judgment aggregation. Finally, Section 5.3 provides, albeit briefly, extra context to the impossibility result touching upon three related topics: the feasibility of non-manipulable aggregation on a class of non-simple agendas; the issue of strategy-proofness in judgment aggregation; the complexity theory of manipulation, and the tempering it offers to the impact of the impossibility of non-manipulable aggregation. The chapter presents and elaborates on results taken mostly from [Die06] and [DL07c].

5.1   TYPES OF MANIPULATION

What do we mean by the manipulation of an aggregation procedure? Or by manipulability of an aggregation function? In this section we clarify the notion of manipulation introducing two of its variants: *agenda manipulation*, whereby an agenda setter can strategically select which issues to let the individuals express themselves on, and *vote manipulation*, whereby individuals themselves can decide to misrepresent their own judgments to the aggregation function in order to enforce a collective outcome that is closer to their own views. While agenda manipulation will be introduced only in passing, vote manipulation will be dealt with in much detail.

5.1.1 AGENDA MANIPULATION

The problem of aggregation starts once an agenda is fixed and the individuals are called to express themselves on the issues in the agenda. Let us for a moment take a step back from this set-up and consider the position of a chairperson entitled to set the agenda of the aggregation problem by selecting the issues to be voted upon. Such a chairperson, if in possession of enough information about the individual opinions in the group, might be able to determine the collective judgment on some of the issues by strategically selecting the issues to appear in the agenda or in a relevant sub-agenda, in the case of rules using only a subset of the available issues (like the premise-based rule). Let us give two different examples of how (sub-)agendas can be manipulated:

Example 5.1 Issue swapping Consider again Example 2.6 and assume the same profile of judgment sets over agenda $A = \pm\{p, p \to q, q\}$: $J_1 = \{p, p \to q, q\}$, $J_2 = \{p, \neg(p \to q), \neg q\}$, $J_3 = \{\neg p, p \to q, \neg q\}$. Suppose the aggregation function is the premise-based rule, where the premises consist in the sub-agenda $A_1 = \pm\{p, p \to q\}$. The collective judgment is therefore $\{p, p \to q, q\}$. A chairperson interested in manipulating the collective judgment toward the rejection of q could now swap q for $p \to q$ in A_1 and use as set of premises the sub-agenda $A_2 = \pm\{p, q\}$:

	p	$p \to q$
J_1	1	1
J_2	1	0
J_3	0	1
J	1	1

$\longmapsto$

	p	q
J_1	1	1
J_2	1	0
J_3	0	0
J	1	0

The rejection of $p \to q$ would then have to be inferred as conclusion, thereby obtaining the collective judgment set $\{p, \neg(p \to q), \neg q\}$.

Example 5.2 Decision framing [CPS08] Now consider the agenda $\pm\{p, q, p \wedge q\}$ and the usual profile: $J_1 = \{p, q, p \wedge q\}$, $J_2 = \{p, \neg q, \neg(p \wedge q)\}$, $J_3 = \{\neg p, q, \neg(p \wedge q)\}$. Assume further that voting proceeds by the premise-based rule on the sub-agenda of premises $\pm\{p, q\}$, which yields the collective judgment $\{p, q, p \wedge q\}$. Now a chairperson willing to get $p \wedge q$ collectively rejected could replace issue q with issue $p \leftrightarrow q$, obtaining the agenda $A' = \{p, p \leftrightarrow q, p \wedge q\}$:

	p	q	$p \wedge q$
J_1	1	1	1
J_2	1	0	0
J_3	0	1	0
J	1	1	1

$\longmapsto$

	p	$p \leftrightarrow q$	$p \wedge q$
J_1	1	1	1
J_2	1	0	0
J_3	0	0	0
J	1	0	0

Premise-based majority would then yield the collective judgment $\{p, \neg(p \leftrightarrow q), \neg(p \wedge q)\}$ which rejects $p \wedge q$.

In the first of the two above examples, the collective judgment on an issue of interest to the chairperson is changed by introducing that issue in the sub-agenda constituting the premises. In fact, since the premise-based rule satisfies a restricted form of independence on its premises[1] the only way to possibly modify the collective judgment on q is clearly to add (resp., remove) it from the set of premises.[2] In the second, the collective judgment is modified by replacing one of the premises by an altogether new complex issue which is either accepted or rejected by each judgment set on the original agenda. This is an instance of the so-called *reframing* of an aggregation problem [CPS08] or *logical* agenda manipulation [Die06]. In general, it is not hard to observe that manipulations of the above types are possible whenever the aggregation rule fails to be independent not only with respect to the formulae of the agenda (as defined in **IND**), but also with respect to all the formulae that are *settled* by the agenda—in the sense that either themselves or their negation follows from each judgment set on that agenda.[3]

5.1.2   VOTE MANIPULATION

In this section we address the form of manipulation that arises when individuals misrepresent their true vote in order to force an outcome of the aggregation process which is closer to their individual views. Let us start again with an example:

Example 5.3   Vote manipulation [DL07c]   Consider once more the $\pm\{p, q, p \land q\}$ agenda and suppose the three individuals are to apply premise-based voting—where the premises are p and q—but are interested solely in the logical conclusions of the aggregation process—namely whether $p \land q$ is the case. Suppose the 'true' judgment sets of the three individuals are $J_1 = \{p, q, p \land q\}$, $J_2 = \{p, \neg q, \neg(p \land q)\}$ and $J_3 = \{\neg p, q, \neg(p \land q)\}$, and suppose furthermore that the first individual votes truthfully, thereby accepting both premises p and q. If the other two individuals are aware of this, how are they going to vote? As they both reject the conclusion $p \land q$, on the basis of the information they have about the first individual, they know that if they both reject both assumptions ($\neg p$ and $\neg q$) they will be able to force their own view through the aggregation modifying the profile as follows:

	p	q	$p \land q$
J_1	1	1	1
J_2	1	0	0
J_3	0	1	0
J	1	1	1

$\longmapsto$

	p	q	$p \land q$
J_1	1	1	1
J_2	0	0	0
J_3	0	0	0
J	0	0	0

---

[1]Whether a premise φ is collectively accepted depends only on the individual votes on φ.

[2]Cf. [Die06].

[3]This condition is known as *strong independence* [Die06]. The reader is referred to this paper and the aforementioned [CPS08] for a more in-depth discussion of agenda manipulation.

In fact, the third individual alone could force outcome $\neg(p \wedge q)$, provided that she knows what the second individual would vote and that she would vote truthfully:

	p	q	$p \wedge q$
J_1	1	1	1
J_2	1	0	0
J_3	0	1	0
J	1	1	1

$\longmapsto$

	p	q	$p \wedge q$
J_1	1	1	1
J_2	1	0	0
J_3	0	0	0
J	1	0	0

The same holds, mutatis mutandis, for the second individual with respect to the third one.

The perspective we have assumed in this example introduces a whole new dimension into judgment aggregation, which has to make with the strategic behavior of individuals. While strategic behavior is the realm of the theory of games [vM44]—and we will briefly touch upon it in Section 5.3.2—in the ensuing sections we will assume a social-choice theoretic perspective on the phenomenon of manipulability, studying it from an axiomatic point of view. In the remainder of this chapter, when referring to manipulability we will mean *vote manipulability* unless explicitly stated otherwise.

5.1.3   MANIPULABILITY: DEFINITION AND CHARACTERIZATION

Example 5.3 above has shown that premise-based aggregation is manipulable: do non-degenerate non-manipulable aggregation rules exist? So let us first introduce manipulability as a property of aggregation functions:

Definition 5.4   Manipulability.   Let $\mathcal{J} = \langle N, A \rangle$ be a judgment aggregation problem. An aggregation function f is:

Manipulable (MAN) iff $\exists P, P' \in \mathbf{P}, \exists i \in N, \exists \varphi \in A$ s.t. $f(P) \neq_\varphi P_i$, $P =_{-i} P'$ and $P_i =_\varphi f(P')$.

I.e., in some profile P collectively rejecting φ, some individual accepting φ is able to force the collective acceptance of φ by submitting a different judgment set P'_i to f, provided all other individuals vote like in P.

A function is said to be non-manipulable (non-**MAN**) otherwise.

Observe that this condition states a mere 'possibility' or 'opportunity' of manipulation. Whether such possibility is attractive or not for the potential manipulator is a different issue having to do with the manipulator's incentives. Non-manipulability is a very strong condition to impose on the aggregation and can be characterized in terms of the properties of independence and monotonicity introduced in Definition 2.18:

Theorem 5.5   Characterizing manipulability [DL07c].   *Let $\mathcal{J} = \langle N, A \rangle$ be a judgment aggregation problem and f an aggregation function. The following assertions are equivalent:*

i) f *does not satisfy* **MAN**;

ii) f *satisfies* **IND** *and* **MON**.

Proof. From (ii) to (i) Assume f satisfies **IND** and **MON** and suppose $\exists \varphi \in X$, $P \in \mathbf{P}$ and $i \in N$ s.t. $P_i \neq_\varphi f(P)$. We will show that $\forall P' \in \mathbf{P}$ s.t. $P' =_{-i} P$ we have that $P_i \neq_\varphi f(P')$, thus proving non-**MAN**. There are two cases: (a) $P_i =_\varphi P_i'$, or (b) $P_i \neq_\varphi P_i'$. As to (a), by **IND** it follows that $f(P') =_\varphi f(P)$, and hence it is still the case that $f(P') \neq_\varphi P_i$. As to (b), since $P_i \neq_\varphi f(P)$, it also follows that $P_i' =_\varphi f(P)$. By **MON**, it follows that $f(P) =_\varphi f(P')$ and hence that $f(P') \neq_\varphi P_i$ as required.

From (i) to (ii) Assume non-**MAN**. (a) We prove that **MON** follows. Take any $\varphi \in X$, $i \in N$ and $P, P' \in \mathbf{P}$ s.t. $P =_{-i} P'$. w.l.o.g. assume that $P_i \not\models \varphi$ and $P_i' \models \varphi$. Now, if $\varphi \in f(P)$, then $P_i \neq_\varphi f(P)$ and by non-**MAN** $f(P) =_\varphi f(P')$. (b) We prove that **IND** follows. Consider any $\varphi \in X$ and $P, P' \in \mathbf{P}$ s.t., $\forall i \in N$: $P_i =_\varphi P_i'$ (the antecedent of **IND**). w.l.o.g. suppose $\varphi \in f(P)$ and assume toward a contradiction that $\varphi \notin f(P')$. It follows that there exists a profile, namely P', such that $P_i' \models \varphi$ but $\varphi \notin f(P')$ and that there exists a profile, namely P, such that $P_i =_\varphi P_i'$ and $\varphi \in f(P)$, thereby implying **MAN**, against the assumption. $\square$

So not only are independence and monotonicity necessary conditions against the manipulability of an aggregation function, but they also suffice to guarantee the non-manipulability of the aggregation. It is instructive to notice that the result holds independently of any assumption on the structure of the agenda and independently of whether the aggregation function is taken to be rational. The theorem can also be interpreted as a characterization of manipulability in terms of sincere and insincere manipulability, to which we turn now.

5.1.4 SINCERE AND INSINCERE MANIPULATION

In the context of preference aggregation, [DvH07] has put forth a refinement of the notion of manipulability in terms of sincere and insincere manipulability. Adapting those insights to the context of judgment aggregation we get to the following definitions:

- f is *insincerely manipulable* iff $\exists P, P' \in \mathbf{P}$, $\exists i \in N$, $\exists \varphi \in A$ s.t. $f(P) \neq_\varphi P_i$, $P =_{-i} P'$, $P_i =_\varphi f(P')$ and $P_i \neq_\varphi P_i'$;

- f is *sincerely manipulable* iff f satisfies **MAN** and $\forall P, P' \in \mathbf{P}, \forall i \in N, \forall \varphi \in A$, if $f(P) \neq_\varphi P_i$, $P =_{-i} P'$, $P_i =_\varphi f(P')$ then $P_i =_\varphi P_i'$.

Intuitively, an aggregation function is insincerely manipulable whenever some individual has the opportunity to bring it about that the collective judgment accepts φ by submitting to the aggregation function a judgment set in which she rejects φ. In other words, the individual lies by rejecting φ while obtaining at the same time that the group accepts φ. Figure 5.1 depicts an example of

	p	q	$(p \wedge q) \leftrightarrow r$	r				p	q	$(p \wedge q) \leftrightarrow r$	r
J_1	1	1	1	1		J_1		1	1	1	1
J_2	1	0	1	0	$\longmapsto$	J_2		1	0	1	0
J_3	0	1	1	0		J_3		0	0	0	1
J	1	1	1	1		J		1	0	1	0

Figure 5.1: Insincere manipulation in the discoursive paradox under premise-based aggregation: p, q and $(p \wedge q) \leftrightarrow r$ are the premises, and r is the conclusion. The third individual (J_3) manages to manipulate the collective judgment to $\neg r$ (her true judgment) by submitting a judgment containing r instead.

such manipulation for premise-based aggregation. Otherwise, an aggregation function is sincerely manipulable whenever it is manipulable but any manipulation about φ is only possible if the individual accepts φ also in her misrepresented judgment set. Example 5.3 has already illustrated cases of manipulation of the sincere kind.

Remark 5.6   Observe that both sincere and insincere manipulability imply manipulability and that if a manipulable function is not sincerely manipulable, it must be insincerely manipulable. So the property of manipulability is equivalent to the disjunction of insincere and sincere manipulability. Closer inspection of the definitions of insincere and sincere manipulability also reveals that insincere manipulability is actually the negation of monotonicity and that sincere manipulability implies the negation of independence (the reader is invited to check these relationships in detail). In the light of these observations Theorem 5.5, acquires a new interesting interpretation and boils down to the more self-evident statement that an aggregation function is not manipulable if and only if it is neither insincerely manipulable, nor sincerely manipulable.

5.2   NON-MANIPULABLE AGGREGATION: IMPOSSIBILITY

Following the line of Chapter 3, we introduce, prove and discuss an impossibility result concerning the existence of non-degenerate aggregation functions which are not manipulable [DL07c]:

> For path-connected agendas, an aggregation function is responsive and non-manipulable if and only if it is a dictatorship.

In other words, when the agenda is path-connected (Definition 2.13) dictatorship is characterized by responsiveness and non-manipulability. It is therefore impossible to aggregate in a non-degenerate way individual judgments into a collective one without at the same time violating responsiveness or introducing the possibility of manipulating behavior by the individuals.

We will prove the theorem by resorting to the same technique we used in Section 3: the ultrafilter method. This allows us to reuse the dictatorship lemma (Lemma 3.6). However, to obtain the desired result we will have to prove a variant of the ultrafilter lemma encountered in Chapter 3 (Lemma 3.5), as this time we are working with weaker aggregation conditions—we do not have systematicity—and with different agenda conditions—we have path-connectedness instead of even-negatability.

5.2.1   AUXILIARY RESULTS

To establish the desired theorem we are going to need four lemmas, which we state and prove in this subsection.

Unanimity   We first show that a responsive and non-manipulable aggregation function is necessarily unanimous:

Lemma 5.7   Unanimity   *If an aggregation function f satisfies* **RES** *and non-***MAN***, then it satisfies* **U**.

Proof. Assume there exists $\varphi \in A$ s.t. $\forall i \in N : P_i \models \varphi$. We proceed to show that $\varphi \in f(P)$. By **RES**, $\exists P'$ s.t. $\varphi \in f(P')$. Recall that, by Theorem 5.5, f satisfies **IND** and **MON**. Take now P' and replace, for all i, P_i' with P_i, thus obtaining P. For each replacement we have two cases: (a) $P_i' \models \varphi$, or (b) $P_i' \models \neg\varphi$. If (a) is the case then, by **IND**, we have that $\varphi \in f(P)$. If (b) is the case then, by **MON**, we also have that $\varphi \in f(P)$. By (a) and (b) we thus conclude that $\varphi \in f(P)$. $\square$

Winning coalitions   We show that non-manipulable and responsive functions on path-connected agendas can be represented by one set of winning coalitions.

Lemma 5.8   Winning coalitions and non-manipulable functions   *Let $\mathcal{J} = \langle N, A \rangle$ be a judgment aggregation problem where A satisfies PC and f be an aggregation function that satisfies non-***MAN** *and* **RES**. *Then there exists a set of winning coalitions $\mathcal{W}$ (Definition 3.3) such that:*

$$\varphi \in f(P) \quad \text{IFF} \quad P_\varphi \in \mathcal{W} \tag{5.1}$$

for all $P \in \mathbf{P}$ and $\varphi \in A$.

Proof. First we show that the set of winning coalitions for each formula in the agenda is the same: (†) $\forall \varphi, \psi \in A : \mathcal{W}_\varphi = \mathcal{W}_\psi$. We split the proof of this claim in two directions: $\mathcal{W}_\psi \supseteq \mathcal{W}_\varphi$ and $\mathcal{W}_\varphi \subseteq \mathcal{W}_\psi$.

$\boxed{\mathcal{W}_\psi \supseteq \mathcal{W}_\varphi}$ Assume $C \in \mathcal{W}_\varphi$. By the assumption of PC (Definition 2.13), we have that $\varphi = \varphi_1 \models_c \ldots \models_c \varphi_k = \psi$ for some $\varphi_1, \ldots, \varphi_k \in A$. We show that, $\forall j : 1 \leq j \leq k, C \in \mathcal{W}_{\varphi_j}$. Proceed by induction.

B: Let $j = 1$, then the claim holds by assumption.

S: Let $1 \leq j < k$, and assume (IH) that $C \in \mathcal{W}_{\varphi_j}$. We prove that $C \in \mathcal{W}_{\varphi_{j+1}}$. Since $\varphi_j \models_c \varphi_{j+1}$ by assumption, there exists $X \subseteq A$ s.t. $X \cup \{\varphi_j, \neg\varphi_{j+1}\}$ is inconsistent but $X \cup \{\varphi_j\}$, $X \cup \{\neg\varphi_{j+1}\}$, $X \cup \{\varphi_j, \varphi_{j+1}\}$ and $X \cup \{\neg\varphi_j, \neg\varphi_{j+1}\}$ are all consistent. Define now a profile P as follows:

$$P_i = \begin{cases} X \cup \{\varphi_j, \varphi_{j+1}\} & \text{if } i \in C \\ X \cup \{\neg\varphi_j, \neg\varphi_{j+1}\} & \text{if } i \in N - C \end{cases}$$

We can observe the following. First, by **U** (Lemma 5.7) we have that $X \subseteq f(P)$. Then, by IH, $C \in \mathcal{W}_{\varphi_j}$ and since $P_{\varphi_j} = C$ we have that $\varphi_j \in f(P)$. Now, since $X \cup \{\varphi_j, \neg\varphi_{j+1}\}$ is inconsistent, we also have that $\varphi_{j+1} \in f(P)$. For any profile P' we therefore have that if $P'_{\varphi_j} = C$ then $P'_{\varphi_{j+1}} = C$. For, suppose not, i.e., $C \notin \mathcal{W}_{\varphi_{j+1}}$. By **IND**, which follows from non-**MAN** by Theorem 5.5, we obtain that for any P', if $P'_{\varphi_{j+1}} = C$ then $\varphi_{j+1} \notin f(P')$. Contradiction.

$\boxed{\mathcal{W}_\varphi \subseteq \mathcal{W}_\psi}$ The proof of this direction is similar and left to the reader.

We can now prove the main claim of the theorem (Formula 5.1). $\boxed{\Leftarrow}$ It holds directly by the above definition of $\mathcal{W}$. $\boxed{\Rightarrow}$ Consider the set of individuals P_φ. For any $P' \in \mathbf{P}$, by **IND**, we have that if $P_\varphi = P'_\varphi$ then $\varphi \in f(P')$. Hence $P_\varphi \in \mathcal{W}_\varphi$ and by (†), $P_\varphi \in \mathcal{W}$. $\qquad\square$

Remark 5.9   Systematicity and non-manipulability The lemma is related to Lemma 3.4, which we encountered in the proofs of the impossibility theorems of Chapter 3. On the one hand, we have that a function is representable in terms of a set of winning coalitions whenever the function is non-manipulable and responsive on a path-connected agenda. On the other hand, Lemma 3.4 showed that if a function is systematic (independently of the type of agenda), then it is representable by a set of winning coalitions, and vice versa. We thus obtain as a corollary that, on path-connected agendas, non-manipulability and responsiveness imply systematicity.

Effects of path-connectedness on the agenda structure   We now show that path-connectedness forces the existence of a specific configuration of consistent and inconsistent sets of formulae of the agenda:

Lemma 5.10   Effects of PC   *Let $\mathcal{J} = \langle N, A \rangle$ be a judgment aggregation problem such that A satisfies PC. There exists an inconsistent set $X \subseteq A$ and three pairwise disjoint sets $Y_1, Y_2, Y_3 \subseteq X$ s.t. $\forall i \in \{1, 2, 3\}$: $(X - Y_i) \cup \neg Y_i$ is consistent, where $\neg Y_i = \{\neg\varphi \mid \varphi \in Y_i\}$.*

Proof. By the assumption of PC it follows that there exists $\varphi = \varphi_1, \ldots, \varphi_n = \neg\varphi$ and $Y'_1, \ldots, Y'_{n-1} \subseteq A$ s.t.:

(i) for $1 \leq i \leq n - 1$, $\{\varphi_i, \neg\varphi_{i+1}\} \cup Y'_i$ is inconsistent;

(ii) for $1 \leq i \leq n - 1$, $\{\varphi_i\} \cup Y'_i$ and $\{\neg\varphi_{i+1}\} \cup Y'_i$ are consistent;

$$
\begin{aligned}
(X - Y_1) \cup \neg Y_1 &= \{\neg(a \succ b), b \succ c, c \succ a\} \\
(X - Y_2) \cup \neg Y_2 &= \{a \succ b, \neg(b \succ c), c \succ a\} \\
(X - Y_3) \cup \neg Y_3 &= \{a \succ b, b \succ c, \neg(c \succ a)\}
\end{aligned}
$$

Figure 5.2: Illustration of Lemma 5.10 on Arrow's agenda. Set $X = \{a \succ b, b \succ c, c \succ a\}$, and $Y_1 = \{a \succ b\}$, $Y_2 = \{b \succ c\}$ and $Y_3 = \{c \succ a\}$.

(iii) for $1 \leq i \leq n - 1$, $\{\varphi_i, \varphi_{i+1}\} \cup Y_i'$ and $\{\neg\varphi_i, \neg\varphi_{i+1}\} \cup Y_i'$ are consistent, which follows from (i) and (ii).

Given these observations we show that (†) $\exists i$ s.t. $1 \leq i \leq n - 1$ and $\{\varphi_i, \neg\varphi_{i+1}\}$ is consistent. Suppose toward a contradiction this is not the case. We then have that all $\{\varphi_1, \neg\varphi_2\} \ldots \{\varphi_{n-1}, \varphi_n\}$ are inconsistent. It follows that all $\{\varphi_1, \varphi_2\}, \{\varphi_1, \varphi_2, \varphi_3\} \ldots \{\varphi_1, \ldots, \varphi_n\}$ are consistent. But $\varphi_1 = \varphi$ and $\varphi_n = \neg\varphi$ by PC. Contradiction.

Having established claim (†), we have that for some i: $\{\varphi_i, \neg\varphi_{i+1}\}$ is consistent and $\{\varphi_i, \neg\varphi_{i+1}\} \cup Y_i'$ is inconsistent by (i), and hence $Y_i' \neq \emptyset$. We can thus prove the desired claim by construction. Define the following sets:

$$
\begin{aligned}
X &= \{\varphi_i, \neg\varphi_{i+1}\} \cup Y_i' \\
Y_1 &= \{\varphi_i\} \\
Y_2 &= \{\neg\varphi_i\}
\end{aligned}
$$

Observe that, by (iii), we have that $(X - Y_1) \cup \neg Y_1$ and $(X - Y_2) \cup \neg Y_2$ are consistent. Now, since X is inconsistent but $\{\varphi_i, \neg\varphi_{i+1}\}$ and Y_i' are both consistent there must exist $Y_3 \subseteq Y_i'$ s.t. $(X - Y_3) \cup \neg Y_3$ is consistent. Finally, Y_1, Y_2 and Y_3 are clearly pairwise disjoint, which concludes the proof. $\qquad\square$

The upshot of this lemma is that, if the agenda is path-connected, then one can find three sets of formulae of the agenda such that: they are all consistent; each two of them agree on at least one formula which is rejected by the third one; all formulae that are accepted by two of them and rejected by the third one form an inconsistent set.

A telling concrete example for this property can be obtained for the Arrow's agenda $\pm\{a \succ b, b \succ c, c \succ a\}$, which we know satisfies PC (recall Example 2.15), and is given in Figure 5.2. This is nothing but a Condorcet cycle, which Lemma 5.10 therefore generalizes to a whole class of agendas.

Ultrafilters of winning coalitions   We are now ready to state the main lemma showing that the set of winning coalitions behaves like an ultrafilter. The reader is encouraged to compare it with Lemma 3.5 of Chapter 3.

Lemma 5.11   Ultrafilter lemma   *Let $\mathcal{J} = \langle N, A \rangle$ be a judgment aggregation problem such that A satisfies PC. Let the aggregation function f satisfy non-**MAN**, **RES** and **RAT**. The set $\mathcal{W}$ is an ultrafilter, i.e.:*

i) *$N \in \mathcal{W}$;*

ii) *if $C \in \mathcal{W}$ then $-C \notin \mathcal{W}$;*

iii) *$\mathcal{W}$ is upward closed: if $C \in \mathcal{W}$ and $C \subseteq C'$ then $C' \in \mathcal{W}$;*

iv) *$\mathcal{W}$ is closed under finite intersections: if $C_1, C_2 \in \mathcal{W}$ then $C_1 \cap C_2 \in \mathcal{W}$.*

Proof. Proofs follow for each of the four claims.

i) The claim follows directly from the fact that f satisfies **U** by Lemma 5.7.

ii) $\boxed{\Rightarrow}$ Suppose toward a contradiction that both $C, -C \in \mathcal{W}$. Consider now a profile P where the judgment sets of the agents in C contain φ and those in $-C$ contain $\neg\varphi$. By Theorem 5.5 f satisfies **IND** from which it follows that $f(P)$ would be inconsistent, against the assumption that f satisfies **RAT**.

$\boxed{\Leftarrow}$ Assume $C \in \mathcal{W}$. Let $C = P_\varphi$ (and therefore $-C = P_{\neg\varphi}$) for some profile P. By Lemma 5.8 we have that $\varphi \in f(P)$ and by **RAT** that $\neg\varphi \notin f(P)$. By the definition of $\mathcal{W}$ (Definition 3.3) we then conclude $-C \notin \mathcal{W}$ as P is such that $-C = P_{\neg\varphi}$ but $\varphi \notin f(P)$.

iii) The claim follows directly from the fact that f satisfies **MON** by Theorem 5.5.

iv) Suppose toward a contradiction that $C_1, C_2 \in \mathcal{W}$ and suppose that $C_1 \cap C_2 \notin \mathcal{W}$. Now put $C' = C_2 - C_1$ and $C'' = N - C_2$. Notice that $C_1 \cap C_2$, $C_2 - C_1$ and $-C_2$ are three disjoint sets covering N. Define now the following profile, which exists by Lemma 5.10:

$$P_i = \begin{cases} (X - Y_1) \cup \neg Y_1 & \text{if } i \in C_1 \cap C_2 = C \\ (X - Y_2) \cup \neg Y_2 & \text{if } i \in C_2 - C_1 = C' \\ (X - Y_3) \cup \neg Y_3 & \text{if } i \in N - C_2 = C'' \end{cases}$$

for X, Y_1, Y_2 and Y_3 defined as in Lemma 5.10.[4] As $C_1 \cap C_2 \notin \mathcal{W}$ by assumption, from ii) it follows that $N - (C_1 \cap C_2) = \overline{C_1} \cup \overline{C_2} \in \mathcal{W}$. Since $C_1 \in \mathcal{W}$ by assumption, it follows by iv) that $C \cup C'' \in \mathcal{W}$. Finally $(C_1 \cap C_2) \cup (C_2 - C_1) = C_2 \in \mathcal{W}$ by assumption. It follows that $f(P) = X$, but X is inconsistent, against the assumption that f satisfies **RAT**.

This completes the proof.   □

---

[4]It might be helpful to use Figure 5.2 for picturing a concrete instance of the claim.

5.2.2   THE IMPOSSIBILITY THEOREM

Although from different assumptions, we can conclude like in Lemma 3.5 that aggregating with non-manipulable, responsive and rational functions on path-connected agendas induces a set of winning coalitions which forms an ultrafilter. Since this ultrafilter is finite, we can again rely on Lemma 3.6 to establish the existence of a dictator, thereby proving the theorem.

Theorem 5.12   Impossibility of non-manipulable aggregation [DL07c].   *Let $\mathcal{J} = \langle N, A \rangle$ be a judgment aggregation problem such that A satisfies PC. An aggregation function f satisfies* **RES**, *non-***MAN** *and* **RAT** *iff it satisfies* **D**.

Proof. $\boxed{\Leftarrow}$ If f satisfies **D** then it trivially satisfies **RES** and non-**MAN**. $\boxed{\Rightarrow}$ By Lemma 5.8, for any $P \in \mathbf{P}$ and $\varphi \in A$:

$$\varphi \in f(P) \quad \text{IFF} \quad P_\varphi \in \mathcal{W}.$$

Then, by Lemmas 5.11 and 3.6 we have that $\{i\} \in \mathcal{W}$ for some $i \in N$ and hence:

$$P_\varphi \in \mathcal{W} \quad \text{IFF} \quad i \in P_\varphi$$

which concludes the proof: $\varphi \in f(P)$ iff $P_i \models \varphi$. $\qquad \square$

The theorem provides a characterization of dictatorship in terms of non-manipulability, responsiveness and collective rationality, on path-connected agendas. As such, notice that its statement is fully analogous to the one of Theorem 3.7, and its proof shares many similarities to the proof of that theorem. Figure 5.3 recapitulates the structure of the proof.

5.3   FURTHER TOPICS: MANIPULATION BEYOND IMPOSSIBILITY RESULTS

Commenting on Theorem 5.12 and its implications, we look at a number of topics and research directions stemming from the issue of non-manipulability in judgment aggregation.

5.3.1   THE POSSIBILITY OF NON-MANIPULABLE AGGREGATION

The proof we have given of Theorem 5.12 relies critically on the path-connectedness assumption for the agenda. As one might expect, on simple agendas non-manipulable aggregation is possible and since propositionwise majority satisfies **MON** and **IND** (Fact 3.1) it also satisfies non-**MAN** (Theorem 5.5).

However, as extensively shown in [Die10], interesting possibility results exist also for richer agendas, and in particular for a large class of so-called *implication agendas* consisting

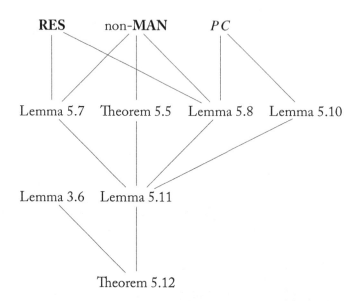

Figure 5.3: Structure of the proof of Theorem 5.12. Edges indicate dependences (from top to bottom) between assumptions, auxiliary results and the impossibility theorem.

.

of implicative connection rules between issues like material implication, but also other non-classical implicative connectives. One such agenda that we have already encountered is the evenly-negatable agenda $\pm\{p, q, p \rightarrow q\}$. Responsive, collectively rational and non-manipulable aggregation on $\pm\{p, q, p \rightarrow q\}$ can for instance be implemented by a quota rule with thresholds $t_p = t_{\neg p} = t_q = t_{\neg q} = \frac{|N|+1}{2}$, $t_{p \rightarrow q} = |N|$ and $t_{\neg(p \rightarrow q)} = 1$. The reader is invited to check that this aggregation function does indeed satisfy **RAT**, **RES**, **IND** and **MON** and hence, by Theorem 5.5, also non-**MAN**.

5.3.2 STRATEGY-PROOF JUDGMENT AGGREGATION

A reader familiar with social choice theory might be tempted to view Theorem 5.12 as a generalization of the result known as Gibbard-Sattherthwaite's theorem,[5] whereby the impossibility of non-manipulability bears not only on preferences, but on path-connected agendas in general. However, the analogy goes only so far as both theorems concern an issue of manipulation of a voting system.

There are two key differences. First, Gibbard-Sattherthwaite's theorem concerns *choice functions*, i.e., functions from profiles of preferences to candidates, while our aggregation functions are instead a generalization of *welfare functions*, i.e., functions from profiles of preferences to

---

[5]Cf. [Gae06, Ch. 5] or [Tay05] for expositions of this result in the context of preference aggregation.

preferences. So Theorem 5.12 should more pertinently be compared with variants of Gibbard-Sattherthwaite's theorem concerning welfare functions, such as the one studied, for instance, in [BS92]. Second, unlike in Gibbard-Sattherthwaite's theorem, our notion of manipulability (Definition 5.4) involves only the possibility of manipulation, and not whether the realization of that possibility would actually be beneficial to the individual, that is, whether the individual has an *incentive* to actually manipulate the aggregation rule. This second difference prompts us to a discussion on the issue of incentive-compatibility or *strategy-proofness* in the context of judgment aggregation, to which we briefly turn.

Incentives in judgment aggregation

Talking about the incentives of individuals means talking about their preferences. Preferences can be modeled in this context as preorders,[6] over the set of all possible judgment sets $\mathbf{J}$. We will denote preferences through the symbol $\succeq$, where $J' \succeq J''$ is taken to mean that judgment set J' is at least as preferred as judgment set J''.

Preferences were not included in the definition of a judgment aggregation problem (Definition 2.1). So two possibilities arise. Either judgment aggregation problems are to be extended with an explicit representation of the preferences of each individual, or they can be built through appropriate stipulations associating a specific preference to each judgment set an individual might truthfully hold.

According to this latter option, the preference of an individual will be a function of her judgment set. Such function would encode how each individual evaluates a possible outcome of the aggregation depending on which judgment set she holds. Let $\mathsf{Pre}(\mathbf{J})$ denote the set of all preorders over $\mathbf{J}$ and call $g : \mathbf{J} \longrightarrow \mathsf{Pre}(\mathbf{J})$ such a function. The property of strategy-proofness of an aggregation function can then be defined as follows:

Definition 5.13 Strategy-proofness. Let $\mathcal{J} = \langle N, A \rangle$ be a judgment aggregation problem, and $g : \mathbf{J} \longrightarrow \mathsf{Pre}(\mathbf{J})$. An aggregation function f is:

Strategy-proof (SP) w.r.t. g iff $\forall i \in N$ and $\forall P, P' \in \mathbf{P}$ s.t. $P =_{-i} P'$: $f(P) \succeq_i f(P')$ where $\succeq_i = g(P_i)$.

I.e., for all individuals and all profiles, submitting the truthful judgment is at least as preferable as misrepresenting it.

Intuitively, f is strategy-proof if no matter what the (g-generated) preferences of individuals are, it is never strictly preferable for them to misrepresent their true judgment set. To put it in game-theoretic terms, it is always a weakly dominant strategy[7] for all the individuals to be truthful toward the aggregation function.

---

[6]We recall that a preorder is a binary relation which is reflexive and transitive. A total preorder is a preorder $\preceq$ which is in addition total, i.e.: $\forall x, y$ either $x \preceq y$ or $y \preceq x$.

[7]See [LBS08, Ch. 3] for the definition of dominant strategy.

Strategy-proofness and non-manipulability

Still, strategy-proofness can be shown to be equivalent to the property of non-manipulability under specific assumptions about the behavior of function g.

Definition 5.14 Closeness of judgment sets. Let $\mathcal{J} = \langle N, A \rangle$ be a judgment aggregation problem and J, J', J'' be judgment sets. We say that J' is at least as close as J'' to J (notation: $J' \sqsupseteq_J J''$) if and only if $\forall \varphi \in A$ if $J =_\varphi J''$ then $J =_\varphi J'$.

Closeness is a ternary relation ordering judgment sets by how 'close' they are respectively to a third judgment set. Intuitively, a judgment set J' is at least as close as J'' w.r.t. J if J' agrees with J on at least all issues on which J'' also agrees with J. It is not difficult to see that the closeness relation $\sqsupseteq$ is a preorder but, note, it is not necessarily a total preorder.

Example 5.15 Closeness Let us consider an example based on the agenda of the discursive dilemma. Let $J = \{p, q, r \leftrightarrow (p \wedge q), r\}$, $J' = \{\neg p, q, r \leftrightarrow (p \wedge q), \neg r\}$ and $J'' = \{\neg p, \neg q, r \leftrightarrow (p \wedge q), \neg r\}$. We have that $J' \sqsupseteq_J J''$ and $J'' \not\sqsupseteq_J J'$. So, J' is strictly closer to J than J''. Consider now $J''' = \{p, \neg q, r \leftrightarrow (p \wedge q), \neg r\}$. We have that $J' \not\sqsupseteq_J J'''$ and $J''' \not\sqsupseteq_J J'$, that is, J' and J''' are incomparable by closeness with respect to J.

Under the assumption that individuals' preferences obey closeness, it becomes possible to show that non-manipulability and strategy-proofness (with respect to a closeness-based notion of preference) define the same condition:

Theorem 5.16 Equivalence of non-MAN and SP [DL07c]. *Let $\mathcal{J} = \langle N, A \rangle$ be a judgment aggregation problem, f an aggregation function and let g be such that $\sqsupseteq_J \subseteq g(J)$. It holds that f satisfies* **SP** *w.r.t. g if and only if it satisfies non-***MAN***.*

Sketch of proof. $\boxed{\Rightarrow}$ Assume that f satisfies **SP** and, toward a contradiction, that f satisfies **MAN**. Then for some $\varphi \in A, i \in N$ and $P, P' \in \mathbf{P}$ such that $P =_{-i} P'$ we have that $P_i \neq_\varphi f(P)$ and $P_i =_\varphi f(P')$. We show that if this is the case, then for some closeness-respecting preference, i would have an incentive to manipulate. Consider the preference $\succeq$ so defined: $J' \succeq_{P_i} J''$ iff $J'' \models \varphi$ implies $J' \models \varphi$.[8] Clearly $\succeq$ is a preorder and it is closeness-respecting. So, for $g(P_i) = \succeq$ we obtain that $f(P') \succeq f(P)$ against the assumption that f satisfies **SP**. $\boxed{\Leftarrow}$ Similar and left to the reader. $\square$

---

[8]Notice that this preference is a total preorder consisting of two classes of equally preferable judgment sets: an upper class of judgment sets agreeing on φ with P_i; and a lower class of all judgment sets disagreeing on φ with P_i. Intuitively, this is the preference of an individual who is concerned solely about item φ.

The key condition here is that the closeness relation $\sqsupseteq_J$ is a subrelation of $g(J)$, that is, the individual's preferences are a refinement or extension of the closeness relation. Preferences of this type are called *closeness-respecting*. So the theorem states that if individuals' preferences are closeness-respecting then the mere existence of a possibility of manipulation is equivalent to the existence of a profitable possibility of manipulation. The theorem can be seen as a reformulation of Theorem 5.12 which is closer in spirit to the Gibbard-Satterthwaite's theorem: a judgment aggregation function satisfies responsiveness and strategy-proofness with respect to closeness-respecting preferences if and only if it is a dictatorship.

It is worth briefly commenting on the intuition backing the notion of closeness-respecting preferences. Closeness assumes that individuals are driven only by the independent acceptance or rejection of individual issues: they prefer outcomes that satisfy more of the issues that they truthfully accept over outcomes that satisfy fewer of such issues. In other words, their preferences are not sensitive to issues being accepted or rejected *in bundles*. For instance, for an individual who truthfully accepts p and q, a closeness-respecting preference would always rank the acceptance of either p or q above the rejection of both. On the other hand, a preference which is not closeness-respecting could rank the rejection of both p and q above the acceptance of only one of the two.[9]

Remark 5.17   Preference, closeness and distance   Distance metrics of the type we have discussed at the end of Chapter 4 are natural candidates to express a notion of preference of individuals over judgment sets [EKM07]. Intuitively, for a generic distance measure d and judgment sets J (the truthful judgment set the individual holds) and J', J'', the individual weakly prefers J' over J'' if and only if $d(J', J) \leq d(J'', J)$. The reader is invited to check that the Hamming distance considered at the end of Chapter 4 generates closeness-respecting preferences.

Judgment aggregation and game theory

The section has just scratched the surface of the sort of strategic issues that might arise in a judgment aggregation setting. More generally, a strategic perspective on voting opens the door to the application of a wealth of notions, results and techniques from game theory [vM44, LBS08] to judgment aggregation. While applications of game theory to the theory of voting in preference aggregation are, to date, extensive, the same cannot be said for judgment aggregation.[10] We consider this one of the promising open research avenues in contemporary judgment aggregation.

5.3.3   COMPLEXITY AS A SAFEGUARD AGAINST MANIPULATION

Whether impossibility results such as Theorem 5.12 should bear real worries about the effective deployment of concrete aggregation procedures is a much debated question. An interesting

---

[9]It is important to realize that preference of this latter sort occur naturally. Think of whether you would prefer to have only one of the two shoes in a pair rather than having none.

[10]For a preliminary contribution in this direction see [GPS09].

perspective on this issue becomes available once one opens up the study of actual aggregation procedures to complexity-theoretic considerations.

Complexity and voting

The application of complexity theory[11] to voting has gained considerable attention in recent years [FHH10]. It is rooted in earlier work in preference aggregation [BTT89, RVW11] and is a currently thriving area of research for both judgment aggregation [EGP12] and, more generally, computational social choice theory [CELM07, BCE13] within artificial intelligence.

The basic intuition behind this body of work is to consider an aggregation function acceptable when it is 'easy' to compute its outcome, and if it is manipulable, when it is 'hard' for an individual to actually manipulate it. The exact meaning of 'easiness' and 'hardness' in this context can be cashed out using complexity theory and is sketched here below.[12] Although we will focus only on the problems of winner determination and of manipulation, the application of complexity theory has moved well beyond those problems considering, in particular, other forms of election control [BEER12] like agenda manipulation, which we have briefly introduced at the beginning of the chapter, and *bribery*, whereby an external agent tries to force a desired result of the aggregation by bribing (within a given budget) some of the individuals to modify their votes.

The winner determination and manipulation problems

In judgment aggregation, the winner determination problem is the problem of checking whether a given formula of the agenda belongs to the output of a given aggregation rule, on a given judgment profile. When it comes to the applicability of an aggregation rule, one would like this problem to be 'computationally easy' to solve. A 'computationally easy' or tractable problem is meant to be a problem that can be solved by an algorithm in time which grows (at worst) polynomially with the size of the input (i.e., the formula, the agenda and the profile).[13] Problems that are not 'easy' to solve are considered to be 'hard' or intractable. This is the case for problems that can be solved by an algorithm in time which grows (at least) exponentially with the size of the input,[14] or problems which at least require their solutions to be checkable by an algorithm in time which grows (at worst) polynomially with the size of the input.[15]

It has been shown in [EGP12] that the winner determination problem for the class of quota rules and for the premise-based rule is 'easy', while it has been shown to be 'hard' for distance-based rules.

The manipulation problem is the problem of checking whether, for a given individual, agenda and judgment profile, there exists a judgment set which, when fed into the input pro-

---

[11]For an accessible introduction to complexity theory we refer the reader to [Gol10]. A standard reference book is [Pap94].

[12]Whether classifications of this type, which are based on standard complexity classes, fit the purpose is currently object of debate in the artificial intelligence literature (cf. [FP10, Wal11]).

[13]This is the class of polynomial time problems.

[14]This is the class of exponential time hard problems.

[15]This latter is the class of non-deterministic polynomial time hard problems. This class is conjectured to be different from the class of polynomial time problems.

file of the aggregation rule, a collective judgment set is obtained which is strictly better for that agent. And by "strictly better" we mean closer according to some specific distance measure, like the Hamming distance (cf. Definition 5.13). In other words, the problem consists in checking whether a given individual can obtain a more favorable outcome by misrepresenting her judgment set, given all other individuals vote in the same way. For the actual deployment of an aggregation rule, one would therefore like its manipulation problem to be 'hard'. This is the case, for instance, for the premise-based rule [EGP12].

CHAPTER 6

Aggregation Rules

From the previous chapters it should be clear that, until recently, the literature on judgment aggregation focused mainly on impossibility theorems and on devising ways to avoid such impossibility results. This is different from what happens in voting theory, where voting rules are defined and studied *per se*. And yet, as it became evident to the researchers gathered at the 2011 workshop "New Developments in Judgement Aggregation and Voting Theory"[1] held in Freudenstadt (Germany), things are starting to change. Several researchers, from within different disciplines, independently began to define concrete aggregation rules for judgment aggregation. This chapter is dedicated to providing a snapshot of this young and ongoing line of research.

Chapter outline: Section 6.1 introduce some general definitions we will be using in the chapter. Section 6.2 to Section 6.4 each defines a specific family of minimization-based rules. The chapter builds on [LPSvdT11, LPSvdT12, NPP11], although it will follow more closely [LPSvdT12], where the family of rules are classified according to a minimization (or maximization) principle. The minimization criterion provides a useful general classification of different rules and, at the same time, it encompasses rules independently defined by other researchers, as we shall note through the chapter. The chapter concludes with some pointers to related literature.

6.1   INTRODUCTION

In Chapter 1 we have seen that the study of voting in preference aggregation was stimulated by practical and specific problems, like the election inside the Academy of Sciences in France. Concrete voting rules were proposed and studied, but it was only with Arrow that the axiomatic approach was applied to social choice theory. Interestingly, judgment aggregation has followed the inverse path. Starting from List and Pettit and, then, with Dietrich and List, the axiomatic method (combined with a logical framework) was employed to investigate abstract judgment aggregation problems, while almost no concrete rule was designed. Exceptions are some of the rules we introduced in Chapter 2 and those we surveyed in Chapter 4, like the premise and conclusion-based procedures, the sequential and quota-based rules and the distance-based ones. Until recently, the literature on judgment aggregation focused mainly on axiomatic studies which led to many impossibility results, that we have explored in Chapter 3 and Chapter 5. As observed in [Die13, Pigng], judgment aggregation has now entered a new phase, one in which the increase of interest in judgment aggregation by researchers from artificial intelligence and multi-agent

---

[1]http://vwll.ets.kit.edu/Workshop_Judgement_Aggregation_and_Voting.php

systems is driving to the definitions and explorations of families of concrete aggregation rules. This chapter presents the initial and ongoing research on this topic.

Minimization

Here we follow [LPSvdT11, LPSvdT12], where several judgment aggregation rules based on minimization are introduced. The motivation for such approach is that several voting rules are based on some minimization (or maximization) principles, and that minimization is also a guiding principle often used in the logic-based knowledge representation community to deal with inconsistency (as in belief revision and belief merging). Yet, with the exception of distance-based rules, minimization had not been exploited to handle inconsistencies arising in judgment aggregation. To overcome this gap, [LPSvdT11, LPSvdT12] introduce four families of minimization-based rules. The inclusion relationships among those rules as well as their social choice-theoretic properties (e.g., unanimity and monotonicity) are also studied. Here we follow their classification and we illustrate those families by few concrete rules.[2] As we will see, interestingly, several rules have been introduced independently by different researchers.

The idea behind the families of rules introduced in [LPSvdT11, LPSvdT12] is to define ways to minimize changes, for example, in the collective judgment set, or in the portion of a profile one needs to remove (or, equivalently, to maximize the portion of a profile one can keep) in order to guarantee a consistent collective judgment set. Each family specifies one particular way such a minimization (resp. maximization) is defined.

Terminology

Let us begin by introducing few general definitions that will be used in the chapter. Recall that given an agenda $A = \pm\{\varphi_1, \ldots, \varphi_m\}$, the *pre-agenda* (or set of issues) associated with A is $[A] = \{\varphi_1, \ldots, \varphi_m\}$. Unless otherwise specified, Γ is the tautology $\top$. In this chapter we call a *sub-agenda* any subset of the agenda. Given a judgment profile P, let us now call the output $f_{maj}(P)$ of propositionwise majority the *majoritarian judgment set* of P, and let us denote it $m(P)$ to ease notation.

The rules we are going to consider in this chapter are irresolute aggregation functions (recall Remark 2.5), that is, their type is $f : \mathbf{P} \longrightarrow \wp(\wp(A))$. Irresoluteness is a natural property to leverage in order to circumvent impossibility results, and in fact we have already encountered instances of irresolute functions in Chapter 4, where we discussed distance-based rules.

We say that a profile is *majority-consistent* iff $m(P)$ is a consistent set of formulae.[3] A judgment aggregation rule f is *majority-preserving* iff, for every majority-consistent profile P, $f(P) = \{m(P)\}$.[4]

---

[2]Here, we consider only three of those four families, as the fourth is the distance-based family of rules, and we have already reviewed some of those rules in Chapter 4.

[3]Cf. the necessary and sufficient domain-restriction condition for majority consistency seen in Section 4.1.2.

[4]The framework of [LPSvdT12] is more general than the one we present here. The conjunction of all logical dependencies among agenda's issues forms one consistent formula $\Gamma \in \mathcal{L}$, called the *constraint*.

	$p \wedge r$	$p \wedge s$	q	$p \wedge q$	t
6 voters	1	1	1	1	1
4 voters	1	1	0	0	1
7 voters	0	0	1	0	0
$m(P)$	1	1	1	0	1

Figure 6.1: A majority-inconsistent profile.

Throughout the chapter we will be using a running example taken from [LPSvdT12] to illustrate the different aggregation rules. Let us introduce it here.

Example 6.1 [LPSvdT12]   Let $[A] = \{p \wedge r, p \wedge s, q, p \wedge q, t\}$ be the pre-agenda and let us suppose the profile P is composed by 17 individuals who express their judgments as in Figure 6.1. The majoritarian set of P is $m(P) = \{p \wedge r, p \wedge s, q, \neg(p \wedge q), t\}$, which is an inconsistent judgment set. Therefore, P is not majority-consistent.

Let us now turn to the first family of minimization-based rules.

6.2   RULES BASED ON THE MAJORITARIAN JUDGMENT SET

One way to ensure a consistent outcome is to calculate $m(P)$ and, when this is not consistent, to restore consistency by minimally removing some issues in the agenda (and thus also the individual and collective judgments on them). These are rules based on the majoritarian judgment set and form the first family of rules defined in [LPSvdT11, LPSvdT12]. Choosing consistent subsets of $m(P)$ that are maximal with respect to set inclusion or cardinality are two natural ways to interpret the minimality criterion.

If we denote by $Max(m(P), \subseteq)$ the set of all maximal consistent subsets of $m(P)$ with respect to set inclusion, and by $Max(m(P), |.|)$ the set of all maximal consistent subsets of $m(P)$ with respect to cardinality, the maximal sub-agenda (MSA) and the maxcard sub-agenda ($MCSA$) rules are defined as follows:

Definition 6.2   Maximal and maxcard sub-agenda rules.   Let $\mathcal{J} = \langle N, A \rangle$ be a judgment aggregation problem and $P \in \mathbf{P}$:

$$
\begin{aligned}
MSA(P) &= Max(m(P), \subseteq) \\
MCSA(P) &= Max(m(P), |.|).
\end{aligned}
$$

	$p \wedge r$	$p \wedge s$	q	$p \wedge q$	t
6 voters	1	1	1	1	1
4 voters	1	1	0	0	1
7 voters	0	0	1	0	0
	1	1	1		1
$MSA(P)$	1	1		0	1
			1	0	1

Figure 6.2: Maximal sub-agenda rule.

	$p \wedge r$	$p \wedge s$	q	$p \wedge q$	t
6 voters	1	1	1	1	1
4 voters	1	1	0	0	1
7 voters	0	0	1	0	0
$MCSA(P)$	1	1	1		1
	1	1		0	1

Figure 6.3: Maxcard sub-agenda rule.

Both *MSA* and *MCSA* are clearly majority-preserving rules.

Example 6.3 [LPSvdT12]   Consider the same agenda and profile *P* of Example 6.1. Figures 6.2 and 6.3 show respectively the results of *MSA(P)* and *MCSA(P)*. When taking the maximal consistent subsets with respect to set inclusion (*MSA(P)*), we see that there are three ways to render the collective judgment set consistent: either we remove $p \wedge q$ from the agenda, or we remove proposition q only, or we remove both $p \wedge r$ and $p \wedge s$. Since the first two options can restore consistency by removing just one proposition each, the third one (that is, eliminating both $p \wedge r$ and $p \wedge s$) will not figure among the outcomes of *MCSA(P)*.

As the example shows, the maximal sub-agenda and the maxcard sub-agenda rules restore consistency by removing some issues from the agenda, and thereby returning collective sets that are possibly neither complete nor deductively closed.

The maxcard sub-agenda coincides with a judgment aggregation rule defined in [MO09], called *Endpoint$_d$*, when d is the Hamming distance. Moreover, independently of Lang et al., *MSA* and *MCSA* have been defined by Nehring et al. [NPP11], who called the maximal sub-agenda rule "Condorcet admissible set," and the maxcard sub-agenda "Slater rule."

6.3   RULES BASED ON THE WEIGHTED MAJORITARIAN JUDGMENT SET

We have seen that rules of the first family aim at restoring consistency by minimally removing items of the agenda. They do not, however, take into account whether an agenda item is supported by many or only few individuals. This information is taken up by the second family of rules introduced in [LPSvdT11].

Let the *weighted majoritarian judgment set* of a profile P be $w(P) = \{\langle \varphi, |P_\varphi|\rangle, \varphi \in A\}$. Intuitively, given a profile P, $w(P)$ records the support received by each agenda item (the support being the number of agents that have that item in their judgment sets). The rules based on the weighted majoritarian judgment set re-establish social consistency by keeping the agenda items that received the highest support. There are two natural ways to achieve this. One is to find the set of all consistent sub-agendas maximizing the total support, that is, each item is associated to the number of individuals who support it and the sub-agendas with the highest total support are selected. The other is to establish an order on the agenda items, from the elements of the agenda supported by the largest majority to those which received fewest votes and, following such order, to accept in the collective set as many issues as it is possible without introducing an inconsistency. These are called respectively the *maxweight sub-agenda rule* and the *ranked sub-agenda rule* [LPSvdT11], and the latter corresponds to the "leximin rule" in [NP11].

In order to illustrate a specific rule of the family, we give the definition and an example of the maxweight sub-agenda rule. For any sub-agenda $A' \subseteq A$, the *weight* of A' with respect to P is defined by $w_P(A') = \sum_{\varphi \in A'} |P_\varphi|$. The maxweight sub-agenda rule is defined as follows:

Definition 6.4   Maxweight sub-agenda rule.   Let $\mathcal{J} = \langle N, A\rangle$ be a judgment aggregation problem. $MWA(P)$ is the set of all consistent sub-agendas maximizing w_P.

So, MWA outputs all consistent subsets of the agendas that maximize $w_P(A')$. It is not difficult to show, and the reader might want to try, that MWA is majority-preserving.

Example 6.5   [LPSvdT12]   Consider the same agenda and profile P of Example 6.1. Let us first calculate the support received by each agenda item in P:

$$w(P) = \{ \quad \langle p \wedge r, 10\rangle, \langle \neg(p \wedge r), 7\rangle,$$
$$\langle p \wedge s, 10\rangle, \langle \neg(p \wedge s), 7\rangle,$$
$$\langle q, 13\rangle, \langle \neg q, 4\rangle,$$
$$\langle p \wedge q, 6\rangle, \langle \neg(p \wedge q), 11\rangle,$$
$$\langle t, 10\rangle, \langle \neg t, 7\rangle \quad \}$$

If we now look at which sub-agendas maximize the total support while returning a consistent collective outcome, we find that there is only one consistent sub-agenda with maximal weight, and this is $A' = \{p \wedge r, p \wedge s, q, p \wedge q, t\}$, with $w_P(A') = 49$. Thus, $MWA(P) = \{\{p \wedge r, p \wedge s, q, p \wedge q, t\}\}$.

MWA is called "Prototype" in [MO09], "median rule" in [NPP11], "simple scoring rule" in [Die13], and can be shown to be equivalent to the distance-based rule $F^{d_H, \Sigma}$ (Formula 4.1) we discussed in Chapter 4 [LPSvdT12, Die13].

The weighted majoritarian judgment set of a profile $w(P)$ corresponds to the definition of "tally vector" in [NP11].[5] The underlying principle of [NP11] is that of *supermajoritarian efficiency*, that states that a supermajority on one proposition can be overruled only when—by doing this—a larger supermajority on another proposition can be recovered in a consistent set. Nehring and Pivato define and characterize the family of "support rules" as the rules based on the weighted majoritarian judgment set. They show that Slater-like rules (a.k.a. *MCSA*) are driven by counting majorities, whereas Leximin-like ones (a.k.a. ranked sub-agenda rules) tend to preserve unanimity. The Median rule (a.k.a. *MWA*) is located in between Slater-like rules and Leximin-like ones.

6.4 RULES BASED ON THE REMOVAL OR CHANGE OF INDIVIDUAL JUDGMENTS

Rather than minimally changing the agenda, the third family of rules considers ways to minimally change a profile in order to obtain a resulting majority-consistent profile. For example, *Young judgment aggregation rule* (so-called for being the judgment aggregation counterpart of the Young voting rule[6]) removes the minimum number of individuals so to obtain a majority-consistent profile.

We will need two auxiliary notion. Given a profile $P = \langle J_i \rangle_{i \in N}$ and a subset of individuals $Q \subseteq N$, the restriction of P to Q is $P_{\downarrow Q} = \langle J_j \rangle_{j \in Q}$, and is called a *sub-profile* of P. Let then $MSP(P)$ be the set of majority-consistent sub-profiles of P of maximal length, i.e., such that they contain a maximal number of individuals. So, for $P' \in MSP(P)$ we have that $m(P')$ is a consistent and complete set, i.e., a judgment set. The Young judgment aggregation rule is defined as follows:

Definition 6.6 Young judgment aggregation rule. Let $\mathcal{J} = \langle N, A \rangle$ be a judgment aggregation problem, and $P \in \mathbf{P}$:

$$Y(P) = \{m(P') \mid P' \in MSP(P)\}.$$

[5]Another family of methods, the *Support-based Aggregation Correspondences* has been recently introduced [KEM13] to guarantee, on the one hand, that the determination of the collective judgment sets takes into account the logical relations among agenda's issues and, on the other hand, the strength of support on each issue.

[6]In voting, Young rule is based on removing voters in order to obtain a Condorcet winner. Young rule outputs an alternative that can be made a Condorcet winner by the least removal of voters.

Rule Y is clearly majority-preserving. Let us illustrate how it works on our running example.

Example 6.7 [LPSvdT12] Consider the same agenda and profile P of Example 6.1. The minimal way to reduce the profile in order to restore consistency is to remove three of the individual judgment sets $\{p \wedge r, p \wedge s, q, p \wedge q, t\}$. Thus, the following judgment set is the result of the Young judgment aggregation rule:

$$Y(P) = \{\{\neg(p \wedge r), \neg(p \wedge s), q, \neg(p \wedge q), t\}\}.$$

Other rules studied in the literature are the *reversed Young rule*, which seeks minimal ways to enlarge the profile by duplicating some of the individual judgment sets, and the *minimal number of atomic changes rule* which looks for a minimum number of switches in the original individual judgment sets to restore consistency [LPSvdT12]. The last rule has been inspired by Dodgson's voting rule that we have seen in Chapter 1, and corresponds to the $Full_d$ rule of [MO09], where d is the Hamming distance.

6.5 FURTHER TOPICS

Some of the rules that we have reviewed in this chapter (most notably the Young rule) have been defined by analogy with well-known voting rules, thus providing judgment aggregation counterparts to some voting rules. The formal connections between some of the rules seen in this chapter and existing voting rules have been established in [LS13]. When studying such relations, it becomes clear that the consistency of the collective judgment could be interpreted as consistency with the transitivity constraint in a preference agenda, or with the constraint stating the existence of at least one undominated alternative. The correspondences between judgment aggregation rules and voting rules are thus determined by requiring the collective set to be consistent with the transitivity constraint or with respect to the existence of undominated alternatives. So, for example, the maxcard sub-agenda rule is proved to be equivalent to the Slater rule when the consistency constraint is set to be the transitivity, and to the Copeland rule when the constraint corresponds to the dominating alternative.

Dietrich [Die13] studied a further class of judgment aggregation rules defined by analogy with well-known voting rules, i.e., scoring rules, like the plurality rule or Borda rule, which we have encountered in Chapter 1. Translated into the framework of judgment aggregation, scoring rules select those collective judgments that maximize the total score. Interestingly, his findings reveal that several existing judgment aggregation rules (e.g., distance-based, premise and conclusion-based) can be re-modeled as scoring rules.

As mentioned at the beginning of the chapter, the definition and study of concrete aggregation rules for judgment aggregation is a line of research in its infancy. Here we provided an overview of recent developments by focusing on specific research papers.

CHAPTER 7

Deliberation

The theory of judgment aggregation presented so far has relied on voting rules, albeit of possibly very different kinds, as its sole method of aggregation. But there exists also a more dynamic and unstructured side of aggregation, the one of group deliberation. Before engaging in a voting process, or even in place of it, we are used to exchange and update opinions, influence one another, and possibly attempt to reach a consensual position. This last chapter of the book is dedicated to bringing the phenomenon of deliberation into focus.

Chapter outline: The chapter discusses two, so far unrelated, formal perspectives on deliberation. The first one, which we will deal with in Section 7.1 concerns an established model of probabilistic opinion change [DG74, LW81]. This model has not yet been fully taken up in the literature on judgment aggregation, and the section will make such connection explicit. The second perspective, which we will discuss in Section 7.2, consists in a preliminary attempt [Lis11] to a formal analysis of deliberation from a traditional judgment aggregation vantage point, conceptualizing it as a process of pre-vote profile transformation. The final section will point to some related work and, in particular, to some open research questions. The chapter is based on material from [Jac08, Ch. 8] and [DG74, Leh76, Bra07] for the first approach, and [Lis11] for the second one. Although not particularly technical, it presupposes some basic knowledge of probability theory.

7.1 DELIBERATION AND OPINION POOLING

In this section we present a simple and influential model of dynamic opinion and consensus formation within a group. The model was developed independently in the statistics and probability theory literature by De Groot [DG74], and in the philosophical literature by Lehrer and Wagner [Leh76, LW81]. The basic abstractions of the model were first introduced and studied by French [Fre56] and Harary [Har59], within the social sciences. We will present the model by recasting it as a judgment aggregation problem.

7.1.1 PROBABILISTIC JUDGMENTS

Suppose a group of individuals needs to find out which event obtains among a set of mutually exclusive and exhaustive events, e.g.: "it rains" or "it does not rain;" "the economy will grow" or "the economy will stagnate" or "the economy will shrink." And suppose each individual holds a belief—in the form of a probability distribution—over those events. Or, similarly, suppose they have to allocate shares of a fixed amount of money to a set of projects, and each one of them holds an opinion about how much to allocate to each project. Problems like these can be viewed as instances of a special class of judgment aggregation problems.

Definition 7.1   Probabilistic agendas, judgments and profiles.   A *probabilistic agenda* is a finite set $A \subseteq At$. A *probabilistic judgment* is a function $J : A \longrightarrow [0, 1]$ s.t.

$$\sum_{p \in A} J(p) = 1.$$

Again, we will denote the set of probabilistic judgments by **J**. A *probabilistic (judgment) profile* is a tuple $P = \langle J_i \rangle_{i \in N} \in \mathbf{J}^{|N|}$ of individual probabilistic judgments.

So a judgment aggregation problem $\mathcal{J} = \langle N, A \rangle$, where A is a probabilistic agenda and where judgments are taken to be probabilistic, describes a general class of collective decision-making problems where the individuals in N need to find a collective probability distribution, or allocation of $[0, 1]$ values, over the elements of A.

7.1.2   A STOCHASTIC MODEL OF DELIBERATION

We now look at deliberation as a dynamic process of update of one's opinions in view of the opinions held by others.

Influence

Suppose now the individuals in the group are ready to revise their own beliefs in view of what other members might believe. How much they care about the opinions of other group members varies—some might be considered especially knowledgeable by somebody, and untrustworthy by somebody else—and can be represented again in a familiar format:

Definition 7.2   Influence judgments and profiles.   Let $\mathcal{J} = \langle N, A \rangle$ be a judgment aggregation problem where A is a probabilistic agenda. An *influence judgment* (w.r.t. $\mathcal{J}$) is a function $T : N \longrightarrow [0, 1]$ s.t.:

$$\sum_{i \in N} T(i) = 1.$$

An *influence (judgment) profile* is a tuple $Q = \langle T_1, \dots, T_{|N|} \rangle$ of individual influence judgments. The ith entry of Q is, as usual, denoted Q_i.

Intuitively, influence judgments quantify how much one individual's judgments may be influenced by the judgments of the other individuals in the group, whatever the sources of such influence are, like power, charisma, trust or respect. An influence profile could rightly be seen as an abstract representation of social influence and power within a group. It can be conveniently represented as a matrix (see below), or as a graph or network where individuals $i, j \in N$ are linked by a directed edge if $Q_i(j) > 0$ (see [Jac08, Ch. 8]).

Changing opinions

Given an initial probabilistic profile P and an influence profile Q, individual iterated revision of probabilistic judgments by influence judgments yields a stream of profiles $P^{(0)}, P^{(1)}, \ldots$ defined inductively as follows: [Base] $P^{(0)} = P$; [Step] $P^{(n+1)} = \left\langle P_1^{(n+1)}, \ldots, P_{|N|}^{(n+1)} \right\rangle$ where $P_i^{(n+1)} = \left\langle P_i^{(n+1)}(p_1), \ldots, P_i^{(n+1)}(p_{|A|}) \right\rangle$ and:

$$P_i^{(n+1)}(p_j) = \sum_{1 \leq k \leq |N|} Q_i(k) P_k^{(n)}(p_j) \qquad (7.1)$$

for $1 \leq i \leq |N|$ and $1 \leq j \leq |A|$.

Formula 7.1 describes what is commonly known in statistics and probability theory as a (linear) *opinion pool* [Sto61]: the next value of a variable, about which the members of the group may disagree, is obtained by combining each current individual value of that variable—i.e., each probabilistic judgment about that variable—weighted by a vector of values—i.e., the influence judgments. This specific update is carried out by each individual at each step. So the stream of probabilistic profiles is the result of distributed linear opinion pooling by each member of the group.

Deliberation as matrix multiplication

Let us move on with two simple observations. Just like judgment profiles (Remark 2.19), probabilistic profiles can be represented by matrices of type $|N| \times |A|$, but where entries are values in the $[0, 1]$ interval rather than in the $\{0, 1\}$ set:

$$\begin{pmatrix} P_1(p_1) & P_1(p_2) & \cdots & P_1(p_{|A|}) \\ P_2(p_1) & P_2(p_2) & \cdots & P_2(p_{|A|}) \\ \vdots & \vdots & \ddots & \vdots \\ P_{|N|}(p_1) & P_{|N|}(p_2) & \cdots & P_{|N|}(p_{|A|}) \end{pmatrix}$$

Likewise, influence profiles can also be conveniently represented by $|N| \times |N|$ matrices like:

$$\begin{pmatrix} Q_1(1) & Q_1(2) & \cdots & Q_1(|N|) \\ Q_2(1) & Q_2(2) & \cdots & Q_2(|N|) \\ \vdots & \vdots & \ddots & \vdots \\ Q_{|N|}(1) & Q_{|N|}(2) & \cdots & Q_{|N|}(|N|) \end{pmatrix}$$

where $Q_i(j)$ encodes i's influence judgment over j, or more simply the influence of j over i. Notice that, intuitively, the diagonal in such matrices reports how much each individual trusts her own beliefs.

Both the above matrices are *stochastic* since each cell has a non-negative value and the values in each row add up to 1. The stream $P^{(0)}, P^{(1)}, \ldots$ can be viewed as the result of the iteration of the multiplication of the matrix of the influence profile Q and the matrix of the probabilistic profile P, i.e.:[1]

$$P^{(n+1)} = Q \cdot P^{(n)} = Q^{n+1} \cdot P^{(0)} \qquad (7.2)$$

where $Q^{n+1} = \underbrace{Q \cdot \ldots \cdot Q}_{n+1 \text{ times}}$ is the $n + 1^{th}$ power of Q. The second equality holds by the associativity of matrix multiplication. So, the $n + 1^{th}$ probabilistic profile is the result of multiplying the $n + 1^{th}$ power of the influence matrix by the initial probabilistic profile.

Two questions arise naturally: Under what conditions does the stream of profiles generated by Formula 7.2 converge to a limit, whereby the process of deliberation modeled by the stream concludes reaching a precise outcome? And what kind of probabilistic profile is obtained at such a limit, that is, what do the individuals believe in the limit?

Convergence—or whether the deliberation ends

Let us start with the first question. We say that the stream $P^{(0)}, P^{(1)}, \ldots$ *converges* if the limit $\lim_{n\to\infty} P^{(n)} = \lim_{n\to\infty} Q^n \cdot P^{(0)}$ exists. Convergence can be characterized based on how sets of individuals influence one another within the influence matrix Q. Let $C \subseteq N$, we say that C is:

strongly connected (w.r.t. Q) if for any two individuals $i, j \in C$ there exists a path $i = i_0, i_1, \ldots, i_m = j$ such that $Q_{i_k}(i_{k+1}) > 0$ for $0 \leq k < m$. I.e., each individual in C is connected, through a directed influence path, to any other individual in that subgroup.

aperiodic (w.r.t. Q) if the greatest common divisor of the lengths of all paths $i = i_0, i_1, \ldots, i_m = j$ where $i = j$ (cycles) is 1.[2] I.e., put roughly, the cycles of influence connecting individuals to themselves are 'out of phase' and do not compare to one another.

closed (w.r.t. Q) if $\nexists i \in C$ and $\nexists j \in N - C$ s.t. $Q_i(j) > 0$. I.e., the individuals in C are not influenced by any individual outside C.

Using these conditions on Q, one can obtain the following characterization of convergence, which we state without proof:[3]

Theorem 7.3 Convergence characterization [GJ10]. *Let $\mathcal{J} = \langle A, N \rangle$ be a judgment aggregation problem with A a probabilistic agenda, and let P be a probabilistic profile and Q an influence*

---

[1]In this case we are multiplying a matrix of type $|N| \times |N|$ (a square matrix) by a matrix of type $|N| \times |A|$. The two matrices are therefore conformable since the number of columns of the first is equal to the number of rows of the second. It follows that the product is a matrix of the latter type, i.e., $|N| \times |A|$. It might be instructive to recall here how matrix multiplication is computed. Let M and M' be two conformable matrices of type $m \times n$ and $n \times r$ respectively. The value of the ij-entry in $M \cdot M'$ is given by $\sum_{1 \leq k \leq n} a_{ik} b_{kj}$ where a_{ik} is the ik-entry of M and b_{kj} is the kj-entry of M'. Applying this formula to the multiplication $Q \cdot P^{(n)}$, the reader will notice that the result is precisely Formula 7.1: $(Q \cdot P^{(n)})_i(p_j) = P_i^{(n+1)}(p_j) = \sum_{1 \leq k \leq |N|} Q_i(k) P_k^{(n)}(p_j)$.

[2]To fix intuitions, it is not the case that N is aperiodic if Q induces a bipartite graph. It is the case that C is aperiodic if at least one individual influences her own opinion by a non-negative value, and there is therefore a cycle of length 1 within C.

[3]The reader is referred to [GJ10] for one.

profile. The stream $P^{(0)}, P^{(1)}, \ldots$ converges if and only if for every $C \subseteq N$, if C is strongly connected and closed w.r.t. Q, then C is aperiodic w.r.t. Q.

Intuitively, a necessary and sufficient condition for the convergence of the deliberative process is that each group of individuals, wherein everybody is connected to everybody else and nobody is influenced by anybody outside the group, is also such that all its internal cycles of influence are 'out of phase'.

Example 7.4   Convergence and non-convergence [Jac08]   Let $N = \{1, 2, 3\}$ and take some initial probabilistic profile P. Consider the two following influence profiles:

$$Q = \begin{pmatrix} 0 & \frac{1}{2} & \frac{1}{2} \\ 1 & 0 & 0 \\ 0 & 1 & 0 \end{pmatrix} \qquad R = \begin{pmatrix} 0 & \frac{1}{2} & \frac{1}{2} \\ 1 & 0 & 0 \\ 1 & 0 & 0 \end{pmatrix}$$

In neither Q nor R the individuals trust their own judgments (all diagonals contain only 0-entries). The graphs induced by Q and R are both rather sparse. In Q individual 1 is influenced by 2 and 3 (with equal weights), 2 is influenced only by 1, and 3 only by 2. The only set of individuals which is strongly connected and closed is, in this case, N itself. We have then two cycles (modulo permutations) within N of length two and, respectively three: $1, 2, 1$ and $1, 3, 2, 1$. N is therefore aperiodic, and $P \cdot Q$ must therefore converge. We have in fact:

$$\lim_{n \to \infty} Q^n \cdot P = \begin{pmatrix} \frac{2}{5} & \frac{2}{5} & \frac{1}{5} \\ \frac{2}{5} & \frac{2}{5} & \frac{1}{5} \\ \frac{2}{5} & \frac{2}{5} & \frac{1}{5} \end{pmatrix} \cdot P$$

where all individuals have the same appraisal of each other's influence: individuals 2 and 3 are, in everybody's view, both twice as influent as individual 3.

The situation depicted by R is slightly different: 1 is again influenced by 2 and 3 (with equal weights), but 2 and 3 are both influenced by 1. Also in this case N is the only strongly connected and closed subgroup. But here we have two cycles of length two: $1, 2, 1$ and $1, 3, 1$. N is therefore not aperiodic and the stream does not converge. In fact, it displays an oscillatory behavior:

$$R^2 = \begin{pmatrix} 1 & 0 & 0 \\ 0 & \frac{1}{2} & \frac{1}{2} \\ 0 & \frac{1}{2} & \frac{1}{2} \end{pmatrix} \quad R^3 = \begin{pmatrix} \frac{1}{2} & \frac{1}{2} & 0 \\ 1 & 0 & 0 \\ 1 & 0 & 0 \end{pmatrix} \quad R^4 = \begin{pmatrix} 1 & 0 & 0 \\ 0 & \frac{1}{2} & \frac{1}{2} \\ 0 & \frac{1}{2} & \frac{1}{2} \end{pmatrix} \quad \cdots$$

Consensus—or whether an ending deliberation is successful

To answer the second question we look at the especially interesting case of when, in the limit, the beliefs of the individuals coincide, that is, when there is unanimity on the values assigned to every issue in the agenda. More precisely, we say that the stream $P^{(0)}, P^{(1)}, \ldots$ *reaches a consensus* if, for $i, j \in N$, $\lim_{n \to \infty} P_i^{(n)} = \lim_{n \to \infty} P_j^{(n)}$. The reaching of consensus can be characterized, using the above properties of sets of individuals as follows:

Theorem 7.5   Consensus characterization [Jac08].   *Let $\mathcal{J} = \langle A, N \rangle$ be a judgment aggregation problem with A a probabilistic agenda. For any probabilistic profile P and influence profile Q, the stream $P^{(0)}, P^{(1)}, \ldots$ reaches a consensus if and only if there exists exactly one $C \subseteq N$ such that C is strongly connected and closed w.r.t. Q and, in addition, C is aperiodic w.r.t. Q.*

Again, we refer the reader to [Jac08, Ch. 8] for a proof of this result. To interpret the theorem, notice that it strengthens Theorem 7.3 by requiring the existence of exactly one set of individuals enjoying the properties that guarantee convergence in the deliberation process. Intuitively, if strongly connected and closed groups of individuals model communities where the patterns of influence are such that probability distributions can align (provided the groups are aperiodic), then the existence of only one of such communities suggests that alignment will happen (if the group is aperiodic) toward one probability distribution, thereby generating consensus. As an illustration, notice that these conditions hold for influence profile Q of Example 7.4: group $\{1, 2, 3\}$ is the only strongly connected and closed set of individuals, and it is aperiodic.

The section has sketched the fundamental ideas behind the stochastic model of opinion change due to De Groot, Wagner and Lehrer. Much more can be said and a vast, and growing, literature exists on more realistic extensions of the model. The reader is referred to [Jac08, Ch. 8] for a more comprehensive treatment from the standpoint of network analysis.

7.1.3   OPINION POOLING AND JUDGMENT AGGREGATION

At the heart of the model we just presented is linear opinion pooling (Formula 7.1), whereby each individual updates her own beliefs at each time step according to the values of the influence matrix. Linear opinion pooling has been shown [LW81, Bra07] to be the only procedure for aggregating probability distributions via a function $f : \mathbf{P} \longrightarrow \mathbf{J}$ that satisfies some suitable variants of the independence property—if $\forall i \in N \colon P_i(p_j) = P_i'(p_j)$ then $f(P)(p_j) = f(P')(p_j)$—and of the unanimity property[4]—if $\forall i \in N \colon P_i(p_j) = 0$ then $f(P)(p_j) = 0$.

Now consider a probabilistic agenda A but assume that judgments, although still satisfying the constraint $\sum_{p \in A} J(p) = 1$ (for $p \in A$), are of the type $J : A \longrightarrow \{0, 1\}$. In other words, they treat the elements of the agenda as mutually exclusive and exhaustive propositions and therefore assign 1 to exactly one of them. Under these constraints, the above characterization implies that linear pooling is possible only with influence matrices containing one individual with maximal

---

[4]The property in question is known as *zero unanimity*.

influence (i.e., 1) on everybody in the group, that is, a dictator. We can then obtain impossibility results through a novel route, the one of pooling (cf. [Bra07]). The reader might find it instructive to check that probabilistic agendas (Definition 7.1), under the above constraints, are actually non-simple and path-connected agendas.

Probabilistic agendas are special cases of richer agendas that can be defined on a probability space and a set of 'measurable' events (a so-called σ-algebra).[5] A rich and interesting literature has addressed the issue of aggregation of probabilities in this more general setting like [Mcc81, Wag82], obtaining limitative results analogous to the impossibility theorems of judgment aggregation we have studied in this book. Formal relationships between these two types of results have recently been object of research in, among others, [DL10a] and [Her13].

7.2   DELIBERATION AS JUDGMENT TRANSFORMATION

An impossibility result we proved in Chapter 3 (Theorem 3.8) showed how, on sufficiently rich agendas, any aggregation function satisfying some mild constraints is necessarily an oligarchy. But among all the possible oligarchies the only acceptable one is, arguably, the one consisting of *all* individuals. In other words, the theorem seems to point to the unanimity rule as the only acceptable form of judgment aggregation. Later, in Chapter 4 (Section 4.1) we have seen that if the individuals hold somewhat 'aligned'—although not unanimous—views, aggregation is possible through (and uniquely through) the propositionwise majority rule.

These results prompt the natural question: if aggregation is possible only when individuals' disagreement is limited, are there ways to reduce individuals' disagreement ahead of voting? After all, the stochastic model presented in the previous section seemed to exemplify precisely a process of disagreement reduction, albeit in a probabilistic setting: convergent deliberation can act, in effect, as an aggregation function yielding unanimous opinions among the group members. In the present section we discuss and formalize this view of deliberation as a process of transformation of individual judgments. In doing so we will leave the probabilistic set up used in the previous section, and get back to the standard theory we have worked with in the rest of the book. But before getting to the formal details, let us briefly comment on the relation between deliberation and voting.

7.2.1   DELIBERATION AND VOTING

Modern political theory recognizes deliberation and voting as the two preeminent collective decision-making processes. The two are sometimes seen as opposite ways of resolving disagreement: in voting, individuals remit themselves to the collective decision taken through some appropriate rule, even when this does not coincide with their personal views; while in deliberation,

---

[5]As already noticed, in our probabilistic agendas all events are mutually exclusive and exhaustive (e.g., elementary or atomic).

individuals seek to convince one another, thereby aligning their individual views toward a consensus.[6]

On another account, voting and deliberation can be viewed as two equally important phases in the structuring of a collective decision-making process. Deliberation typically precedes voting, informing members of the group of the individual views of other members and of their reasons. It therefore 'prepares' the individual opinions of each individual for a subsequent aggregation via voting. And in some accounts it is claimed that this 'preparation' for the act of voting offers a way of circumventing the impossibilities of aggregation.[7]

7.2.2 JUDGMENT TRANSFORMATION FUNCTIONS

We are going to study deliberation as a function transforming judgment profiles:

Definition 7.6 Transformation function.   Let $\mathcal{J} = \langle N, A \rangle$ be a judgment aggregation problem. A *transformation function* for $\mathcal{J}$ is a function $t : \mathbf{P} \longrightarrow \wp(A)^n$. For an individual i and profile P, $t(P)_i$ denotes the ith set in the transformed profile $t(P)$.

A transformation function takes as input a judgment profile and outputs a tuple of sets of formulae.[8] By the above definition a transformation function accepts as input any possible profile but does not necessarily output a profile of judgment sets. Intuitively, the process modeled by a transformation function maps the individual judgments of each individual to a new individual opinion which is neither necessarily consistent nor necessarily complete.

Before introducing a few examples, let us state the following important observation:

Remark 7.7 Transformation as composition of aggregation   It is important to observe that a transformation function t can equivalently be viewed as a tuple $\langle f_i \rangle_{i \in N}$ where each f_i is an aggregation function. Intuitively, transforming a profile can be viewed as the combined process of

---

[6]Cf. the following quote from [Els86]:

> The core of the theory [of deliberation] [...] is that rather than aggregating or filtering preferences, the political system should be set up with a view to changing them by public debate and confrontation. The input to the social choice mechanism would then not be the raw, quite possibly selfish or irrational, preferences [...] but informed and other-regarding preferences. Or rather, there would not be any need for an aggregation mechanism, since a rational discussion would tend to produce unanimous preferences. [Els86, p. 112]

[7]Cf. the following quote from [DL03]:

> Deliberation facilitates pursuit of several escape-routes from the impossibility results invoked by social-choice theoretic critics of democracy [...]. Thus social choice theory shows exactly what deliberation must accomplish in order to render collective decision making tractable and meaningful, suggesting that democracy must in the end have a deliberative aspect. [DL03, p. 27]

[8]Notice that, modulo the probabilistic setup, this is precisely the type of the operation of multiplication by an influence matrix we have encountered in Section 7.1. Recall in particular Formula 7.2.

each individual internally aggregating the initial profile—i.e., the view of each other individual—to obtain a new set of judgments.[9]

7.2.3   EXAMPLES OF TRANSFORMATION FUNCTIONS

By way of example, let us introduce two simple instances of transformation functions. Like in the case of aggregation functions, we will refer to instances of transformation functions as transformation rules.

Deference to majority:

$$t(P)_i \quad = \quad f_{maj}(P) \tag{7.3}$$

for all $i \in N$, where $f_{maj}(P)$ is the propositionwise majority function. I.e., as result of the transformation, each individual assumes as individual judgment set the collective set that would be obtained by applying the majority rule.

Opinion leader:

$$t(P) \quad = \quad \langle P_{o(i)} \rangle_{i \in N} \tag{7.4}$$

where $o : N \longrightarrow N$ is a function assigning to each individual an *opinion leader* whose judgment set the individual assumes in the transformed profile.[10] I.e., each individual changes her judgment set into the judgment set of some individual (possibly herself[11]).

These are two very simple forms of judgment transformation modeling some trivial forms of deliberation: in the case of deference to majority the deliberation can be seen merely as a process of declaration of vote intentions whereby the majority position is unanimously internalized by each individual; in the opinion leader case deliberation consists in blindly following the vote intention of a selected individual.

Other more complex transformation rules can clearly be defined,[12] but in this section we will concern ourselves primarily with providing a first charting of the boundaries between possible and impossible transformation rules. In doing this we provide yet another application of the axiomatic method we first introduced in Chapter 3.

7.3   LIMITS OF JUDGMENT TRANSFORMATION

The section presents a simple impossibility result concerning the transformation of judgment profiles. The impossibility will be obtained from known impossibility results for aggregation functions.

---

[9]Notice how this is not dissimilar from each individual carrying out a linear opinion pool in the stochastic model described in the previous section.

[10]Alternatively, one can see o as an influence matrix with values ranging in $\{0, 1\}$.

[11]The identity function is the special opinion leader rule where each individual assumes herself as opinion leader.

[12]The reader is encouraged to consult [Lis11].

7.3.1 CONDITIONS ON TRANSFORMATION FUNCTIONS

The conditions echo some of the mapping conditions for aggregation functions we studied in Chapter 2.

Definition 7.8 Conditions on transformation functions.   Let $\mathcal{J} = \langle N, A \rangle$ be a judgment aggregation problem. A transformation function t for $\mathcal{J}$ is:

Rational ($\mathbf{RAT'}$) iff $\forall i \in N, t(P)_i$ is consistent and complete.

I.e., the codomain of t is the set of profiles of judgment sets over $\mathcal{J}$.

Unanimous ($\mathbf{U'}$) iff $\forall P \in \mathbf{P}, \forall \varphi \in A :$ IF $\forall i \in N, P_i \models \varphi$ THEN $\forall i \in N, \varphi \in t(P)_i$.

I.e., if all individuals agree on a formula of the agenda, then the transformation preserves the agreement on that formula.

Responsive ($\mathbf{RES'}$) iff $\forall i \in N, \exists P, P' \in \mathbf{P} : P =_{-i} P', P_i \neq P'_i$ AND $t(P)_i \neq t(P')_i$.

I.e., for any individual there are always at least two profiles, which are identical except for that individual's judgment set, such that the judgment set of that individual remains different also in the two transformed profiles.

Independent ($\mathbf{IND'}$) iff $\forall \varphi \in A, \forall P, P' \in \mathbf{P} :$ IF $[\forall i \in N : P_i \models \varphi$ IFF $P'_i \models \varphi]$ THEN $[\forall i \in N : \varphi \in t(P)_i$ IFF $\varphi \in t(P')_i]$.

I.e., if all individuals in two different profiles agree on the acceptance or rejection pattern of a formula, each individual's judgment set in the transformed profiles also does.

Systematic ($\mathbf{SYS'}$) iff $\forall \varphi, \psi \in A, \forall P, P' \in \mathbf{P} :$ IF $[\forall i \in N : P_i \models \varphi$ IFF $P'_i \models \psi]$ THEN $[\forall i \in N : \varphi \in t(P)_i$ IFF $\psi \in t(P')_i]$.

I.e., if all individuals in two different profiles agree on the acceptance or rejection pattern of two formulae (φ is accepted iff ψ is accepted), each individual's judgment set in the transformed profiles also does.

The conditions express constraints on how deliberation is supposed to modify judgment sets: $\mathbf{RAT'}$ imposes that judgment profiles are transformed into judgment profiles, that is, individuals remain 'rational' across the transformation; $\mathbf{U'}$ imposes that an existing unanimity be preserved by the transformation; $\mathbf{RES'}$ imposes that individual do not always change their judgments; $\mathbf{IND'}$ imposes that each issue be deliberated upon independently of any other; finally, $\mathbf{SYS'}$ imposes that issues that are treated equivalently by the individuals in two profiles, will be treated equivalently also in the transformed profiles.

Some of these may be considered fairly uncontroversial constraints, like $\mathbf{RAT'}$, $\mathbf{U'}$ and $\mathbf{RES'}$. Others might appear rather more stringent from the point of view of a deliberative setting, like $\mathbf{IND'}$ and $\mathbf{SYS'}$. Still, they may be viewed as some forms of guarantee that issues are treated in the same way across different deliberations. Whether constraints of this type are sensible for the modeling of specific pre-vote deliberation processes is a question we will not delve into here,[13] and we rather aim now to illustrate how they can be used to explore the space of possible judgment transformation functions.

7.3.2   AN IMPOSSIBILITY RESULT

Given the above conditions on transformation functions we can obtain simple impossibility results like the following:

Theorem 7.9 [Lis11].   *Let $\mathcal{J} = \langle N, A \rangle$ be a judgment aggregation problem such that A satisfies NS and EN, and let t be a transformation function: t satisfies, **RAT′**, **U′**, **RES′** and **SYS′** iff t is the identity function.*

Proof. Recall first of all Remark 7.7: $t = \langle f_i \rangle_{i \in N}$. $\boxed{\Leftarrow}$ If t is the identity function, each f_i is a dictatorship by individual i. As a consequence, t clearly satisfies **U′**, **RES′** and **SYS′**. $\boxed{\Rightarrow}$ From the assumptions of **RAT′**, **U′** and **SYS′** we conclude that each f_i satisfies **RAT**, **U** and **SYS**. By Theorem 3.7, we then obtain that each f_i is a dictatorship by some individual. Take one f_i and suppose, toward a contradiction that such dictator is an individual $j \neq i$. But this contradicts **RES′**, as no two profiles P and P' such that $P =_i P'$ would then generate profiles that differ in the ith judgment set. It follows that each f_i is the dictatorship of i. Hence t is the identity function. $\qquad\square$

The theorem is therefore a fairly direct consequence of Theorem 3.7. In general similar results can be obtained for transformation functions based on corresponding impossibility results for aggregation functions. For example, recalling the impossibility results listed in Section 3.4.1, a similar result can be obtained by weakening **SYS′** to **IND′** and strengthening non-simplicity to path-connectedness.

Intuitively, the theorem sets some precise boundaries about what kind of deliberative processes—intended as transformations of individual views—can reasonably be expected. However, unlike in the case of the aggregation of judgments, conditions like independence and systematicity can be more easily criticized. In aggregation, independence and systematicity guarantee the non-manipulability of the aggregation process (recall Chapter 5), but it is unclear whether non-manipulability is as strong a desideratum from the point of view of a theory of deliberation. A manipulation can, after all, be interpreted as genuine opinion change, and opinion change is precisely what deliberation aims at.

For these reasons, [Lis11] theorizes the possibility of non-independent transformation rules satisfying the property of generating majority-consistent profiles—i.e., profiles that can be consistently aggregated through the propositionwise majority rule.[14] The property is referred to as *cohesion generation*.

---

[13]The reader is referred to [Lis11] for some further comments on this issue.

[14]A typical example are profiles that are unidimensionally aligned (Section 4.1).

7.4 FURTHER TOPICS AND OPEN ISSUES

Deliberation is a much less clear-cut phenomenon than voting, and a formal approach to it, in particular from the point of view of aggregation, is still in its infancy. Deliberation remains an understudied topic in the field of (judgment as well as preference) aggregation. The judgment transformation perspective offers one angle to tackle it formally, but by no means an exhaustive one. Relations between this approach and the ones based on opinion pooling, or others we have not dealt with here (e.g., based on the process of best response dynamics [BCMP13]), are still to be fully understood. Key issues remain open about the exact nature of the deliberative process (what is it the individuals deliberate about?[15]), the incentives that should motivate individuals in changing their opinions leading to a vote (why should an individual modify her individual judgment set?) and strategic considerations that would naturally ensue (why and how can individuals influence one another during deliberation?). Preliminary work toward an answer to the first question is presented in [DL13b], where a model of how rational choice depends on reasons for preferences is introduced. The latter two questions have begun to be addressed by works in game theory, such as [GR01, HL07], which have focused on the strategic issues involved in persuading one another during deliberation, and the informational gain that this involves for the group.

---

[15]For a first interesting attempt to providing an answer to this question see [DL13b].

Bibliography

[Arr50] K. Arrow. A difficulty in the concept of social welfare. *Journal of Political Economy*, 58(4):328–346, 1950. DOI: 10.1086/256963. 7, 35, 48

[Arr63] K. Arrow. *Social Choice and Individual Values*. John Wiley, New York, 2nd edition, 1963. 6, 7, 48, 56

[AvdHW11] T. Ågotnes, W. van der Hoek, and M. Wooldridge. On the logic of preference and judgment aggregation. *Autonomous Agents and Multi-Agent Systems*, 22:4–30, 2011. DOI: 10.1007/s10458-009-9115-8. 14

[BBM81] P. Batteau, J.-M. Blin, and B. Monjardet. Stability of aggregation procedures, ultrafilters, and simple games. *Econometrica*, 49(2):527–534, 1981. DOI: 10.2307/1913328. 47

[BCE13] F. Brandt, V. Conitzer, and U. Endriss. Computational social choice. In G. Weiss, editor, *Multiagent Systems*, pages 213–284. MIT Press, 2013. 13, 88

[BCEF] R. Briggs, F. Cariani, K. Easwaran, and B. Fitelson. Individual coherence and group coherence. In J. Lackey, editor, *Essays in Collective Epistemology*. Oxford University Press. To appear. 68

[BCMP13] S. Brânzei, I. Caragiannis, J. Morgenstern, and A. Procaccia. How bad is selfish voting? In *Proceedings of AAAI'13*, 2013. 110

[BDL+98] S. Benferhat, D. Dubois, J. Lang, H. Prade, A. Saffiotti, and P. Smets. A general approach for inconsistency handling and merging information in prioritized knowledge bases. In *Proceedings of the 6th International Conference on Principles of Knowledge Representation and Reasoning (KR98)*, pages 466–477, 1998. 67

[BdV01] P. Blackburn, M. de Rijke, and Y. Venema. *Modal Logic*. Cambridge University Press, Cambridge, 2001. 33

[BEER12] D. Baumeister, G. Erdélyi, O. Erdélyi, and J. Rothe. Control in judgment aggregation. In *Proceedings of the 6th European Starting AI Researcher Symposium (STAIRS'12)*, pages 22–34. IOS Press, 2012. DOI: 10.3233/978-1-61499-096-3-23. 88

[Ben76] J. Bentham. *A Fragment on Government*. Oxford University Press Warehouse, 1776. 2

[Ber38] A. Bergson. A reformulation of certain aspects of welfare economics. *Quarterly Journal of Economics*, 51(1):310–334, February 1938. DOI: 10.2307/1881737. 7

[Ber66] A. Bergson. *Essays in Normative Economics*. Cambridge, MA: Harvard University Press, 1966. 7

[BK12] I. Beg and A. Khalid. Belief aggregation in fuzzy framework. *The Journal of Fuzzy Mathematics*, 20(4):911–924, 2012. 66

[BKM91] C. Baral, S. Kraus, and J. Minker. Combining multiple knowledge bases. *IEEE Transactions on Knowledge and Data Engineering*, 3(2):208–220, 1991. DOI: 10.1109/69.88001. 67

[BKMS92] C. Baral, S. Kraus, J. Minker, and V. Subrahmanian. Combining multiple knowledge bases consisting of first order theories. *Computational Intelligence*, 8:45–71, 1992. DOI: 10.1111/j.1467-8640.1992.tb00337.x. 66

[BKS07] S. Brams, D. Kilgour, and R. Sanver. A minimax procedure for electing committees. *Public Choice*, 132(3-4):401–420, 2007. DOI: 10.1007/s11127-007-9165-x. 37, 66, 69

[BKZ98] S. Brams, D. Kilgour, and W. Zwicker. The paradox of multiple elections. *Social Choice and Welfare*, 15(2):211–236, 1998. DOI: 10.1007/s003550050101. 71

[Bla48] D. Black. On the rationale of group decision making. *The Journal of Political Economy*, 56:23–34, 1948. DOI: 10.1086/256633. 54, 55

[Bla57] J. H. Blau. The existence of social welfare functions. *Econometrica*, 25(2):302–313, April 1957. DOI: 10.2307/1910256. 8

[Bla58] D. Black. *The Theory of Committees and Elections*. Cambridge University Press, 1958. 1, 2, 4

[Bor84] J.-C. de Borda. Mémoire sur les élections au scrutin. In *Mémoires de l'Académie Royale des Sciences année 1781*, pages 657–665. Imprimerie Royale, Paris, 1784. 3

[BR06] L. Bovens and W. Rabinowicz. Democratic answers to complex questions. An epistemic perspective. *Synthese*, 150:131–153, 2006. DOI: 10.1007/s11229-006-0005-1. 12, 62

[Bra07] R. Bradley. Reaching a consensus. *Social Choice and Welfare*, 29:609–632, 2007. DOI: 10.1007/s00355-007-0247-y. 99, 104, 105

[BS92] W. Bossert and T. Storcken. Strategy-proofness of social welfare functions: The use of the Kemeny distance between preference orderings. *Social Choice and Welfare*, 9:345–360, 1992. DOI: 10.1007/BF00182575. 85

[BTT89] J. Bartholdi, C. Tovey, and M. Trick. The computational difficulty of manipulating an election. *Social Choice and Welfare*, 6(3):227–241, 1989. DOI: 10.1007/BF00295861. 88

[Car11] F. Cariani. Judgment aggregation. *Philosophy Compass*, 6(1):22–32, 2011. DOI: 10.1111/j.1747-9991.2010.00366.x. xvi

[Car37] H. Cartan. Filtres et ultrafiltres. *Comptes Rendus de l'Académie des Sciences*, pages 777–779, 1937. 39

[CELM07] Y. Chevaleyre, U. Endriss, J. Lang, and N. Maudet. A short introduction to computational social choice. In *Proc. SOFSEM 2007: Theory and Practice of Computer Science, Lecture Notes in Computer Science Volume 4362*, pages 51–69. Springer-Verlag, 2007. DOI: 10.1007/978-3-540-69507-3_4. 13, 88

[CF86] J. Coleman and J. Ferejohn. Democracy and social choice. *Ethics*, 97(1):6–25, 1986. DOI: 10.1086/292814. 5

[CGMH+94] S. Chawathe, H. Garcia Molina, J. Hammer, K. Ireland, Y. Papakonstantinou, J. Ullman, and J. Widom. The TSIMMIS project: Integration of heterogeneous information sources. In *Proceedings of IPSJ Conference*, pages 7–18, 1994. 66

[Cha98] B. Chapman. More easily done than said: Rules, reason and rational social choice. *Oxford Journal of Legal Studies*, 18:293–329, 1998. DOI: 10.1093/ojls/18.2.293. 8

[Cha02] B Chapman. Rational aggregation. *Politics, Philosophy & Economics*, 1(3):337–354, Sep 2002. DOI: 10.1177/1470594X02001003004. 61

[Coh86] J. Cohen. An epistemic conception of democracy. *Ethics*, 97(1):26–38, 1986. DOI: 10.1086/292815. 5

[Con85] Marquis de Condorcet, M.J.A.N. de C. *Essai sur l'Application de l'Analyse à la Probabilité des Décisions Rendues à la Pluralité des Voix*. Imprimerie Royale, Paris, 1785. 4

[CP11] M. Caminada and G. Pigozzi. On judgment aggregation in abstract argumentation. *Autonomous Agents and Multi-Agent Systems*, 22(1):64–102, 2011. 54, 71

[CPP11] M. Caminada, G. Pigozzi, and M. Podlaszewski. Manipulation in group argument evaluation. In *Proceedings of the 22nd International Joint Conference on Artificial Intelligence (IJCAI 2011)*, pages 121–126, 2011. DOI: 10.5591/978-1-57735-516-8/IJCAI11-032. 54, 71

[CPS08] F. Cariani, M. Pauly, and J. Snyder. Decision framing in judgment aggregation. *Synthese*, 163(1):1–24, 2008. DOI: 10.1007/s11229-008-9306-x. 74, 75

[Dan10] T. Daniëls. Social choice and the logic of simple games. *Journal of Logic and Computation*, 21(6):883–906, 2010. DOI: 10.1093/logcom/exq027. 47

[Deb54] G. Debreu. Representation of a preference ordering by a numerical function. In R. M. Thrall, C. H. Coombs, and R. L. Davis, editors, *Decision Processes*, pages 159–165. John Wiley, 1954. 52

[dF74] B. de Finetti. *The Theory of Probability*. Wiley, 1974. 68

[DG74] M. De Groot. Reaching a consensus. *Journal of the American Statistical Association*, 69(345):118–121, 1974. 99

[DH10a] E. Dokow and R. Holzman. Aggregation of binary evaluations. *Journal of Economic Theory*, 145(2):495–511, 2010. DOI: 10.1016/j.jet.2009.10.015. 24, 32, 49

[DH10b] E. Dokow and R. Holzman. Aggregation of binary evaluations with abstentions. *Journal of Economic Theory*, 145(2):544–561, 2010. DOI: 10.1016/j.jet.2009.10.015. 32, 59

[Die06] F. Dietrich. Judgment aggregation: (im)possibility theorems. *Economic Theory*, 126:286–298, 2006. DOI: 10.1016/j.jet.2004.10.002. 60, 61, 73, 75

[Die07] F. Dietrich. A generalised model of judgment aggregation. *Social Choice and Welfare*, 28(4):529–565, 2007. DOI: 10.1007/s00355-006-0187-y. 33

[Die10] F. Dietrich. The possibility of judgment aggregation on agendas with subjunctive implications. *Journal of Economic Theory*, 145:603–638, 2010. DOI: 10.1016/j.jet.2007.11.003. 21, 83

[Die13] F. Dietrich. Scoring rules for judgment aggregation. *Social Choice and Welfare*, pages 1–39, 2013. DOI: 10.1007/s00355-013-0757-8. 91, 96, 97

[DKNS01] C. Dwork, R. Kumar, M. Naor, and D. Sivakumar. Rank aggregation methods for the web. In *Proceedings of the 10th international conference on World Wide Web*, WWW '01, pages 613–622. ACM, 2001. DOI: 10.1145/371920.372165. 13

[DL03] J. Dryzek and C. List. Social choice theory and deliberative democracy: A reconciliation. *British Journal of Political Science*, 33:1–28, 2003. DOI: 10.1017/S0007123403000012. 106

[DL07a] F. Dietrich and C. List. Arrow's theorem in judgment aggregation. *Social Choice and Welfare*, 29(1):19–33, 2007. DOI: 10.1007/s00355-006-0196-x. 12, 15, 20, 24, 35, 38, 44, 47, 49, 52

[DL07b] F. Dietrich and C. List. Judgment aggregation by quota rules: Majority voting generalized. *Journal of Theoretical Politics*, 19:391–424, 2007. DOI: 10.1177/0951629807080775. 20, 54, 60, 63, 64

[DL07c] F. Dietrich and C. List. Strategy-proof judgment aggregation. *Economics and Philosophy*, 23:269–300, 2007. DOI: 10.1017/S0266267107001496. 15, 60, 73, 75, 76, 78, 83, 86

[DL08] F Dietrich and C List. Judgment aggregation without full rationality. *Social Choice and Welfare*, 31(1):15–39, 2008. DOI: 10.1007/s00355-007-0260-1. 35, 44, 45, 46, 59

[DL10a] F. Dietrich and C. List. The aggregation of propositional attitudes: Towards a general theory. In *Oxford Studies in Epistemology*, volume 3, pages 215–234. Oxford University Press, 2010. 105

[DL10b] F. Dietrich and C. List. Majority voting on restricted domains. *Journal of Economic Theory*, 145(2):512–543, 2010. DOI: 10.1016/j.jet.2010.01.003. 22, 37, 57, 67

[DL13a] F. Dietrich and C. List. Propositionwise judgment aggregation: The general case. *Social Choice and Welfare*, 40:1067–1095, 2013. DOI: 10.1007/s00355-012-0661-7. 23, 24, 27, 32

[DL13b] F. Dietrich and C. List. A reason-based theory of rational choice. *Nous*, 47(1):104–134, 2013. DOI: 10.1111/j.1468-0068.2011.00840.x. 110

[DM10] F. Dietrich and P. Mongin. The premiss-based approach to judgment aggregation. *Journal of Economic Theory*, 145(2):562 – 582, 2010. DOI: 10.1016/j.jet.2010.01.011. 20, 47, 54, 63

[Dod73] C. L. Dodgson. *A discussion of the various methods of procedure in conducting elections.* Imprint by E. B. Gardner, E. Pickard Hall and J. H. Stacey, Printers to the University, Oxford, 1873. 6

[Dod74] C. L. Dodgson. *Suggestions as to the best method of taking votes, where more than two issues are to be voted on*. Imprint by E. Pickard Hall and J. H. Stacey, Printers to the University, Oxford, 1874. 6

[Dod76] C. L. Dodgson. *A method of taking votes on more than two issues*. Oxford: Clarendon Press, 1876. 6

[DP12] C. Duddy and A. Piggins. A measure of distance between judgment sets. *Social Choice and Welfare*, 39:855–867, 2012. DOI: 10.1007/s00355-011-0565-y. 69

[Dun95] P. M. Dung. On the acceptability of arguments and its fundamental role in nonmonotonic reasoning, logic programming and n-person games. *Artificial Intelligence*, 77:321–357, 1995. DOI: 10.1016/0004-3702(94)00041-X. 70

[DvH07] K. Dowding and M. van Hees. In praise of manipulation. *British Journal of Political Science*, 38:1–15, 2007. DOI: 10.1017/S000712340800001X. 61, 77

[EGP12] U. Endriss, U. Grandi, and D. Porello. Complexity of judgment aggregation. *Journal of Artificial Intelligence Research*, 45:481–514, 2012. DOI: 10.1613/jair.3708. 14, 88, 89

[EK07] D. Eckert and C. Klamler. How puzzling is judgment aggregation? Antipodality in distance-based aggregation rules. *Working paper. University of Graz*, pages 1–7, 2007. 70

[EKM07] P. Everaere, S. Konieczny, and P. Marquis. The strategy-proofness landscape of merging. *Journal of Artificial Intelligence Research*, 28:49–105, 2007. DOI: 10.1613/jair.2034. 87

[Els86] J. Elster. The market and the forum. In J. Elster and A. Hylland, editors, *Foundations of Social Choice Theory*, pages 103–132. Cambridge University Press, 1986. 106

[Els13] J. Elster. Excessive ambitions (ii). *Capitalism and Society*, 8(1), 2013. 8

[EM05] D. Eckert and J. Mitlöhner. Logical representation and merging of preference information. *Multidisciplinary IJCAI-05 Workshop on Advances in Preference Handling*, pages 85–87, 2005. 69

[EP05] D. Eckert and G. Pigozzi. Belief merging, judgment aggregation, and some links with social choice theory. In *Dagstuhl Seminar Proceedings 05321*, 2005. 69

[ERS99] A. Elmagarmid, M. Rusinliewicz, and A. Sheth, editors. *Management of Heterogeneous and Autonomous Database Systems*. Morgan Kaufmann, 1999. 66

[Fey04] M. Fey. May's theorem with an infinite population. *Social Choice and Welfare*, 23:275–293, 2004. DOI: 10.1007/s00355-003-0264-4. 51

[FHH10] P. Faliszewski, E. Hemaspaandra, and L. Hemaspaandra. Using complexity to protect elections. *Communications of the ACM*, 53(11):74–82, 2010. DOI: 10.1145/1839676.1839696. 88

[Fis70] P. C. Fishburn. Arrow's impossibility theorem: Concise proof and infinite voters. *Journal of Economic Theory*, 2(1):103–106, 1970. DOI: 10.1016/0022-0531(70)90015-3. 47, 50

[FP10] P. Faliszewski and A. Procaccia. AI's war on manipulation: Are we winning? *AI Magazine*, 31(4):53–64, 2010. 88

[Fre56] J. French. A formal theory of social power. *Psychological Review*, 63:181–194, 1956. DOI: 10.1037/h0046123. 99

[Gae01] W. Gaertner. *Domain Conditions in Social Choice Theory*. Cambridge University Press, 2001. DOI: 10.1017/CBO9780511492303. 68

[Gae06] W. Gaertner. *A Primer in Social Choice Theory*. Oxford University Press, 2006. 57, 84

[Gär06] P. Gärdenfors. A representation theorem for voting with logical consequences. *Economics and Philosophy*, 22:181–190, 2006. DOI: 10.1017/S026626710600085X. 44, 47, 54, 58, 59

[GE13] U. Grandi and U. Endriss. Lifting integrity constraints in binary aggregation. *Artificial Intelligence*, 199–200:45–66, 2013. DOI: 10.1016/j.artint.2013.05.001. 33

[Gib73] A. Gibbard. Manipulation of voting schemes: A general result. *Econometrica*, 41(4):587–601, Jul. 1973. DOI: 10.2307/1914083. 8

[GJ10] B. Golub and M. O. Jackson. Naive learning in social networks and the wisdom of the crowds. *American Economic Journal: Microeconomics*, 2(1):112–149, 2010. DOI: 10.1257/mic.2.1.112. 102

[GOF83] B. Grofman, G. Owen, and S. L. Feld. Thirteen theorems in search of the truth. *Theory and Decision*, 15(3):261–278, 1983. DOI: 10.1007/BF00125672. 5

[Gol10] O. Goldreich. *P, NP, and NP-completeness*. Cambridge University Press, 2010. DOI: 10.1017/CBO9780511761355. 88

[Got07] S. Gottwald. Many-valued logics. In D. Jacquette, editor, *Handbook of the Philosophy of Sciences*, volume 5. North-Holland, 2007. 33

[GP12] D. Grossi and G. Pigozzi. Introduction to judgment aggregation. In N. Bezhanishvili and V. Goranko, editors, *Lecture Notes on Logic and Computation. ESSLLI'10 and ESSLLI'11, Selected Lecture Notes*, volume 7388 of *LNCS*, pages 160–209. Springer, 2012. DOI: 10.1007/978-3-642-31485-8. xviii

[GPS09] D. Grossi, G. Pigozzi, and M. Slavkovik. White manipulation in judgment aggregation. Proceedings of BNAIC 2009 - The 21st Benelux Conference on Artificial Intelligence, 2009. 87

[GR01] J. Glazer and A. Rubinstein. Debates and decisions: On a rationale of argumentation rules. *Games and Economic Behavior*, 36(2):158–173, 2001. DOI: 10.1006/game.2000.0824. 110

[Gro09] D. Grossi. Unifying preference and judgment aggregation. In P. Decker and J. Sichman, editors, *Proceedings of the 8th International Conference on Autonomous Agents and Multi-Agent Systems (AAMAS 2009)*, pages 217–224. ACM Press, 2009. 52

[Gro10] D. Grossi. Correspondences in the theory of aggregation. In G. Bonanno, B. Loewe, and W. van der Hoek, editors, *Logic and the Foundation of Game and Decision Theory—LOFT 08, Revised and Selected Papers*, volume 6006 of *LNAI*, pages 34–60. Springer, 2010. DOI: 10.1007/978-3-642-15164-4. 52

[Gui52] G. T. Guilbaud. Les théories de l'intérêt général et le problème logique de l'agrégation. *Economie appliquée*, 5:501–505, 1952. 8

[H01] R. Hähnle. Advanced many-valued logics. In D. M. Gabbay and F. Guenthner, editors, *Handbook of Philosophical Logic, 2nd Edition*, volume 2, pages 297–395. Kluwer, 2001. 52

[Han76] B. Hansson. The existence of group preference functions. *Public Choice*, 28:89–98, 1976. DOI: 10.1007/BF01718460. 47, 50

[Har59] F. Harary. A criterion for unanimity in French's theory of social power. In D. Cartwright, editor, *Studies in Social Power*, pages 168–182. Oxford University Press, 1959. 99

[HE09] F. Herzberg and D. Eckert. General aggregation problems and social structure: A model-theoretic generalization of the Kirman-Sondermann correspondence. Working papers, Institute of Mathematical Economics, University of Bielefeld, 2009. 48

[HE12] F. Herzberg and D. Eckert. Impossibility results for infinite-electorate abstract aggregation rules. *Journal of Philosophical Logic*, 41:273–286, 2012. DOI: 10.1007/s10992-011-9203-5. 47, 50

[Her08] F. Herzberg. Judgment aggregation functions and ultraproducts. Working papers, Institute of Mathematical Economics, University of Bielefeld, 2008. 47, 48

[Her10] F. Herzberg. Judgment aggregators and boolean algebra homomorphisms. *Journal of Mathematical Economics*, 46(1):132–140, 2010. DOI: 10.1016/j.jmateco.2009.06.002. 47, 48

[Her12] F. Herzberg. The model-theoretic approach to aggregation: Impossibility results for finite and infinite electorates. *Mathematical Social Sciences*, 64(1):41–47, 2012. DOI: 10.1016/j.mathsocsci.2011.08.004. 47, 48, 50

[Her13] F. Herzberg. Universal algebra for general aggregation theory: Many-valued propositional attitude aggregators as mv-homomorphisms. *Journal of Logic and Computation*, 2013. DOI: 10.1093/logcom/ext009. 47, 105

[HL07] C. Hafer and D. Landa. Deliberation as self-discovery and institutions for political speech. *Journal of Theoretical Politics*, 19(3):329–360, 2007. DOI: 10.1177/0951629807077573. 110

[Hod97] W. Hodges. *A Shorter Model Theory*. Cambridge University Press, 1997. 48

[HPS10] S. Hartmann, G. Pigozzi, and J. Sprenger. Reliable methods of judgement aggregation. *Journal of Logic and Computation*, 20(2):603–617, 2010. DOI: 10.1093/logcom/exp079. 12, 66

[HS11] S. Hartmann and J. Sprenger. Judgment aggregation and the problem of tracking the truth. *Synthese*, 187(1):209–221, 2011. DOI: 10.1007/s11229-011-0031-5. 66

[Jac08] M. O. Jackson. *Social and Economic Networks*. Princeton University Press, 2008. 99, 100, 103, 104

[Joy09] J. Joyce. Accuracy and coherence: Prospects for an alethic epistemology of partial belief. In F. Huber and C. Schmidt-Petriy, editors, *Degrees of Belief*, pages 263–297. Springer, 2009. DOI: 10.1007/978-1-4020-9198-8. 68

[KE09] C. Klamler and D. Eckert. A simple ultrafilter proof for an impossibility theorem in judgment aggregation. *Economics Bulletin*, 29(1):319–327, 2009. 35, 47

[Kem59] J. Kemeny. Mathematics without numbers. *Daedalus*, 88:577–591, 1959. 69

[KEM13] S. Konieczny P. Everaere and P. Marquis. Support-based correspondences for judgment aggregation. In *MFI 2013*, 2013. 96

[KG06] S. Konieczny and E. Grégoire. Logic-based approaches to information fusion. *Information Fusion*, 7:4–18, 2006. DOI: 10.1016/j.inffus.2005.07.002. 65, 67

[KLM04] S. Konieczny, J. Lang, and P. Marquis. DA^2 merging operators. *Artificial Intelligence*, 157:49–79, 2004. DOI: 10.1016/j.artint.2004.04.008. 65, 67, 69

[Koo60] T. C. Koopmans. Stationary ordinal utility and impatience. *Econometrica*, 28:287–309, 1960. DOI: 10.2307/1907722. 50

[Kor92] L.A. Kornhauser. Modeling collegial courts. II. Legal doctrine. *Journal of Law, Economics, and Organization*, 8(3):441–470, 1992. 1, 8, 12

[KPP98] S. Konieczny and R. Pino-Pérez. On the logic of merging. In *Proceedings of the 6th International Conference on Principles of Knowledge Representation and Reasoning*, 1998. 67

[KPP99] S. Konieczny and R. Pino-Pérez. Merging with integrity constraints. *Fifth European Conference on Symbolic and Quantitative Approaches to Reasoning with Uncertainty (ECSQARU'99)*, 7:233–244, 1999. DOI: 10.1007/3-540-48747-6_22. 54, 65, 67, 69

[KPP02a] S. Konieczny and R. Pino-Pérez. Merging information under constraints: A logical framework. *Journal of Logic and Computation*, 12:773–808, 2002. DOI: 10.1093/logcom/12.5.773. 67

[KPP02b] S. Konieczny and R. Pino Pérez. On the frontier between arbitration and majority. In *Proceedings of the 8th International Conference on Principles of Knowledge Representation and Reasoning*, pages 109–118, 2002. 65, 67

[KS72] A. P. Kirman and D. Sondermann. Arrow's theorem, many agents, and invisible dictators. *Journal of Economic Theory*, 5(2):267–277, 1972. DOI: 10.1016/0022-0531(72)90106-8. 47, 50

[KS86] L.A. Kornhauser and L.G. Sager. Unpacking the court. *Yale Law Journal*, 96:82–117, 1986. DOI: 10.2307/796436. 8, 12

[KS93] L.A. Kornhauser and L.G. Sager. The one and the many: Adjudication in collegial courts. *California Law Review*, 81:1–51, 1993. DOI: 10.2307/3480783. 1, 8, 9, 61

[LBS08] K. Leyton-Brown and Y. Shoham. *Essentials of Game Theory*. Morgan & Claypool, 2008. DOI: 10.2200/S00108ED1V01Y200802AIM003. 47, 85, 87

[Leh76] K. Lehrer. When rational disagreement is impossible. *Noûs*, 10(3):327–332, 1976. DOI: 10.2307/2214612. 99

[Lis02] C. List. A possibility theorem on aggregation over multiple interconnected propositions. *Mathematical Social Sciences*, 45(1):1–13, Oct 2002. DOI: 10.1016/S0165-4896(02)00089-6. 54, 56

[Lis04] C. List. A model of path-dependence in decisions over multiple propositions. *American Political Science Review*, 98(3):495 – 513, 2004. DOI: 10.1017/S0003055404001303. 63

[Lis05a] C List. Corrigendum to "A possibility theorem on aggregation over multiple interconnected propositions". *Mathematical Social Sciences*, 52:109–110, Sep 2005. DOI: 10.1016/j.mathsocsci.2005.12.002. 56

[Lis05b] C. List. The probability of inconsistencies in complex collective decisions. *Social Choice and Welfare*, 24(1):3–32, 2005. DOI: 10.1007/s00355-003-0253-7. 62

[Lis11] C. List. Group communication and the transformation of judgments: An impossibility result. *The Journal of Political Philosophy*, 19(1):1–27, 2011. DOI: 10.1111/j.1467-9760.2010.00369.x. 99, 107, 109

[Lis12] C. List. The theory of judgment aggregation: An introductory review. *Synthese*, 187(1):179–207, 2012. DOI: 10.1007/s11229-011-0025-3. xvi, 24, 28, 35, 49, 53, 58

[Lisce] C. List. Social choice theory. In E. N. Zalta, editor, *The Stanford Encyclopedia of Philosophy*, volume Winter 2013 Edition, 2013, http://plato.stanford.edu/archives/win2013/entries/social-choice/. 13

[LM99] J. Lin and A. Mendelzon. Knowledge base merging by majority. In *Dynamic Worlds: From the Frame Problem to Knowledge Management*, pages 195–218. Kluwer, 1999. DOI: 10.1007/978-94-017-1317-7_6. 67

[LP02] C. List and P. Pettit. Aggregating sets of judgments: An impossibility result. *Economics and Philosophy*, 18:89–110, 2002. 15, 32, 44, 61

[LP04] C. List and P. Pettit. Aggregating sets of judgments: Two impossibility results compared. *Synthese*, 140(1):207–235, 2004. DOI: 10.1023/B:SYNT.0000029950.50517.59. 10, 12

[LP09] C. List and C. Puppe. Judgment aggregation: A survey. In *Oxford Handbook of Rational and Social Choice*. Oxford University Press, 2009. DOI: 10.1093/acprof:oso/9780199290420.003.0020. xvi

[LPSvdT11] J. Lang, G. Pigozzi, M. Slavkovik, and L. van der Torre. Judgment aggregation rules based on minimization. In *TARK*, pages 238–246, 2011. DOI: 10.1145/2000378.2000407. 91, 92, 93, 95

[LPSvdT12] J. Lang, G. Pigozzi, M. Slavkovik, and L. van der Torre. Judgment aggregation rules based on minimization - extended version. In *Technical report, Université Paris-Dauphine*, 2012. 91, 92, 93, 94, 95, 96, 97

[LS95] P. Liberatore and M. Schaerf. Arbitration: A commutative operator for belief revision. In *Proceedings of the Second World Conference on the Fundamentals of Artificial Intelligence (WOCFAI '95)*, pages 217–228, 1995. 67

[LS13] J. Lang and M. Slavkovik. Judgment aggregation rules and voting rules. In *Proceedings of the 3rd International Conference on Algorithmic Decision Theory (ADT 2013)*, volume 8176, pages 230–244, 2013. DOI: 10.1007/978-3-642-41575-3_18. 97

[LvL95] L. Lauwers and L. van Liederke. Ultraproducts and aggregation. *Journal of Mathematical Economics*, 24(3):217–237, 1995. DOI: 10.1016/0304-4068(94)00684-3. 47

[LW81] K. Lehrer and C. Wagner. *Rational Consensus in Science and Society*. Springer, 1981. DOI: 10.1007/978-94-009-8520-9. 99, 104

[Mac03] G. Mackie. *Democracy Defended*. Cambridge: Cambridge University Press, 2003. DOI: 10.1017/CBO9780511490293. 13

[May52] K. May. A set of independent necessary and sufficient conditions for simple majority decision. *Econometrica*, 20:680–684, 1952. DOI: 10.2307/1907651. 37

[Mcc81] K. Mcconway. Marginalization and linear opinion pools. *Journal of the American Statistical Association*, 76(374):410–414, 1981. DOI: 10.1080/01621459.1981.10477661. 105

[McL90] I. McLean. The Borda and Condorcet principles: Three medieval applications. *Social Choice and Welfare*, 7:99–108, 1990. DOI: 10.1007/BF01560577. 2

[MD10] P. Mongin and F. Dietrich. Un bilan interprétatif de la théorie de l'agrégation logique. *Revue d'économie politique*, 120(6):929–972, 2010. 10

[MO09] M.K. Miller and D. Osherson. Methods for distance-based judgment aggregation. *Social Choice and Welfare*, 32(4):575–601, 2009. DOI: 10.1007/s00355-008-0340-x. 69, 94, 96, 97

[Mon05] B. Monjardet. Social choice theory and the "Centre de Mathématique Sociale": Some historical notes. *Social Choice and Welfare*, 25(2-3):433–456, 2005. DOI: 10.1007/s00355-005-0012-z. 8

[Mon08] P. Mongin. Factoring out the impossibility of logical aggregation. *Journal of Economic Theory*, 141(1):100–113, 2008. DOI: 10.1016/j.jet.2007.11.001. 61

[Mon11] P. Mongin. Judgment aggregation. In S.O. Hansson and V.F. Hendricks, editors, *The Handbook of Formal Philosophy*. Springer, 2011. xvi, 10

[Mou80] H. Moulin. On strategy-proofness and single peakedness. *Public Choice*, 35(4):437–455, 1980. DOI: 10.1007/BF00128122. 56

[Nas03] J. R. Nash. A context-sensitive voting protocol paradigm for multimember courts. *Stanford Law Review*, 56(1):75–159, 2003. 9

[Neh03] K. Nehring. Arrow's theorem as a corollary. *Economics Letters*, 80(3):379–382, 2003. DOI: 10.1016/S0165-1765(03)00118-6. 49

[Neh05] K. Nehring. The (im)possibility of a Paretian rational. Economics working papers, Institute for Advanced Study, School of Social Science, Nov 2005. 61, 62, 63

[NP02] K. Nehring and C. Puppe. Strategy-proof social choice on single-peaked domains: Possibility, impossibility and the space between. Working Paper, University of California at Davis, 2002. 24

[NP06] K. Nehring and C. Puppe. Consistent judgement aggregation: The truth-functional case. *Social Choice and Welfare*, 31(1):41–57, June 2006. DOI: 10.1007/s00355-007-0261-0. 20, 44, 61

[NP07] K. Nehring and C. Puppe. The structure of strategy-proof social choice. Part I: General characterization and possibility results on median spaces. *Journal of Economic Theory*, 135(1):269–305, 2007. DOI: 10.1016/j.jet.2006.04.008. 22

[NP10a] K. Nehring and C. Puppe. Abstract Arrovian aggregation. *Journal of Economic Theory*, 145(2):467–494, 2010. DOI: 10.1016/j.jet.2010.01.010. 32, 49, 53

[NP10b] K. Nehring and C. Puppe. Justifiable group choice. *Journal of Economic Theory*, 145(2):583–602, 2010. DOI: 10.1016/j.jet.2009.12.004. 61

[NP11] K. Nehring and M. Pivato. Majority rule in the absence of a majority. *Working paper*, 2011. 95, 96

[NPP11] K. Nehring, M. Pivato, and C. Puppe. Condorcet admissibility: Indeterminacy and path-dependence under majority voting on interconnected decisions. In *http://mpra.ub.uni-muenchen.de/32434/*, 2011. 91, 94, 96

[Nur10] H. Nurmi. Voting theory. In D. Rios Insua and S. French, editors, *e-Democracy*, volume 5 of *Advances in Group Decision and Negotiation*, pages 101–123. Springer Netherlands, 2010. DOI: 10.1007/978-90-481-9045-4. 13

[NW68] R. G. Niemi and H. F. Weisberg. A mathematical solution for the probability of the paradox of voting. *Behavioral Science*, 13(4):317–323, 1968. DOI: 10.1002/bs.3830130406. 13

[Odi00] P. Odifreddi. Ultrafilters, dictators and gods. In C. Calude and G. Păun, editors, *Finite Versus Infinite. Contributions to an Eternal Dilemma*, pages 239–246. Springer, 2000. DOI: 10.1007/978-1-4471-0751-4. 35

[Pacds] E. Pacuit. Voting methods. In E. N. Zalta, editor, *The Stanford Encyclopedia of Philosophy*, volume Winter 2012 Edition, 2012, http://plato.stanford.edu/archives/win2012/entries/voting-methods/. 13

[Pap94] C. Papadimitriou. *Computational Complexity*. Addison-Wesley, 1994. 88

[Pat71] P. K. Pattanaik. *Voting and Collective Choice*. London, Cambridge University Press, 1971. 8

[Pet01] P. Pettit. Deliberative democracy and the discursive dilemma. *Philosophical Issues*, 11:268–299, 2001. DOI: 10.1111/j.1758-2237.2001.tb00047.x. 9

[PHG00] D. M. Pennock, E. Horvitz, and C. L. Giles. Social choice theory and recommender systems: Analysis of the axiomatic foundations of collaborative filtering. In *Proceedings of the 17th National Conference on Artificial Intelligence*, pages 729–734. AAAI Press, 2000. 13

[Pig06] G. Pigozzi. Belief merging and the discursive dilemma: An argument-based account to paradoxes of judgment aggregation. *Synthese*, 152:285–298, 2006. DOI: 10.1007/s11229-006-9063-7. 14, 54, 65

[Pigng] G. Pigozzi. The logic of group decisions: Judgment aggregation. *Journal of Philosophical Logic*, forthcoming. 91

[PS04] E. Pacuit and S. Salame. Majority logic. In *Proceedings of the 9th International Conference on the Principles of Knowledge Representation and Reasoning (KR'04)*, pages 598–605, 2004. 51

[PvH06] M. Pauly and M. van Hees. Logical constraints on judgment aggregation. *Journal of Philosophical Logic*, 35:569–585, 2006. DOI: 10.1007/s10992-005-9011-x. 31

[Rev93] P. Revesz.   On the semantics of theory change: Arbitration between old and new information.   In *Proceedings of the 12th ACM SIGACT-SIGMOD-SIGART Symposium on Principles of Databases*, pages 71–82, 1993. DOI: 10.1145/153850.153857. 67

[RGMT06] M. Regenwetter, B. Grofman, A. A. J. Marley, and I. Tsetlin. *Behavioral Social Choice*. Cambridge University Press, 2006. 13, 58, 68

[Rik82] W. H. Riker. *Liberalism Against Populism. A Confrontation Between the Theory of Democracy and the Theory of Social Choice*. San Francisco: W. H. Freeman, 1982. 8

[RL08] I. Rahwan and K. Larson. Welfare properties of argumentation-based semantics. In *Proceedings of the 2nd International Workshop on Computational Social Choice (COMSOC)*, 2008. 71

[Rob38] L. Robbins. Interpersonal comparisons of utility: A comment. *Economic Journal*, 48(192):635–641, 1938. DOI: 10.2307/2225051. 2, 6

[Rou62] J.-J. Rousseau. *Du Contrat Social ou Principes du Droit Politique*. 1762. 5

[RT10] I. Rahwan and F. Tohmé. Collective argument evaluation as judgment aggregation. In *Proc. of 9th AAMAS*, pages 417–424, 2010. 71

[RVW11] F. Rossi, B. Venable, and T. Walsh. *A Short Introduction to Preferences: Between Artificial Intelligence and Social Choice*. Morgan & Claypool, 2011. DOI: 10.2200/S00372ED1V01Y201107AIM014. 13, 88

[Sam47] P. Samuelson. *Foundations of Economic Analysis*. Cambridge, MA: Harvard University Press, 1947. 7

[Sat75] M. A. Satterthwaite. Strategy-proofness and Arrow's conditions: Existence and correspondence theorems for voting procedures and social welfare functions. *Journal of Economic Theory*, 10(2):187–217, April 1975. DOI: 10.1016/0022-0531(75)90050-2. 8

[Sen66] A. Sen. A possibility theorem on majority decisions. *Econometrica*, 34:491–499, Feb 1966. DOI: 10.2307/1909947. 57

[Sen69] A. Sen. Quasi-transitivity, rational choice and collective decisions. *The Review of Economic Studies*, 36(3):381–393, July 1969. DOI: 10.2307/2296434. 8

[Sen70] A. Sen. The impossibility of a Paretian liberal. *Journal of Political Economy*, 78(1):152–157, Jan. - Feb. 1970. DOI: 10.1086/259614. 8

[Sen86] A. Sen. Social choice theory. In K. J. Arrow and M. Intriligator, editors, *Handbook of Mathematical Economics, Vol. III*. North Holland, Amsterdam, 1986. 1, 2, 7

[Sen95] A. Sen. Rationality and social choice. *American Economic Review*, 85(1):1–24, March 1995. 6

[Sen99] A. Sen. The possibility of social choice. *The American Economic Review*, 89(3):349–378, 1999. DOI: 10.1257/aer.89.3.349. 1, 53

[SP69] A. Sen and P. K. Pattanaik. Necessary and sufficient conditions for rational choice under majority decision. *Journal of Economic Theory*, 1(2):178–202, 1969. DOI: 10.1016/0022-0531(69)90020-9. 56

[Spe09] H. Spector. The right to a constitutional jury. *Legisprudence*, 3(1):111–123, 2009. 8

[Sto61] M. Stone. The opinion pool. *Annals of Mathematical Statistics*, 32:1339–1342, 1961. DOI: 10.1214/aoms/1177704873. 101

[Sup05] P. Suppes. The pre-history of Kenneth Arrow's social choice and individual values. *Social Choice and Welfare*, 25:319–326, 2005. DOI: 10.1007/s00355-005-0006-x. 7

[Tar30] A. Tarski. Une contribution à la théorie de la mesure. *Fundamenta Mathematicae*, 15:42–50, 1930. 51

[Tar83] A. Tarski. On some fundamental concepts of metamathematics. In J. Corcoran, editor, *Logic, Semantics, Metamathematics*, pages 30–37. Hackett, 1983. 33

[Tay05] A. D. Taylor. *Social Choice and the Mathematics of Manipulation*. Cambridge University Press, 2005. DOI: 10.1017/CBO9780511614316. 84

[Tul91] G Tullock. Duncan Black: The founding father (23 May 1908–14 January 1991). *Public Choice*, 71(3):125–128, 1991. DOI: 10.1007/BF00155731. 2

[Vac21] R. Vacca. Opinioni individuali e deliberazioni collettive. *Rivista Internazionale di Filosofia del Diritto*, 52:52–59, 1921. 8

[vD80] D. van Dalen. *Logic and Structure*. Springer, 1980. DOI: 10.1007/978-3-662-08402-1. 15

[vM44] J. von Neumann and O. Morgenstern. *Theory of Games and Economic Behavior*. Princeton University Press, 1944. 76, 87

[Wag82] C. Wagner. Allocation, Lehrer models, and the consensus of probabilities. *Theory and Decision*, 14:207–220, 1982. DOI: 10.1007/BF00133978. 105

[Wal11] T. Walsh. Where are the hard manipulation problems? *Journal of Artificial Intelligence Research*, 42:1–29, 2011. DOI: 10.1613/jair.3223. 88

[Wil75] R. Wilson. On the theory of aggregation. *Journal of Economic Theory*, 10(1):89–99, 1975. DOI: 10.1016/0022-0531(75)90062-9. 32

[Wil09] J. Williamson. Aggregating judgements by merging evidence. *Journal of Logic and Computation*, 19(3):461–473, June 2009. DOI: 10.1093/logcom/exn011. 66

[YL78] H. P. Young and A. Levenglick. A consistent extension of Condorcet's election principle. *SIAM Journal of Applied Mathematics*, 35:285–300, 1978. DOI: 10.1137/0135023. 69

[You74] H. P. Young. An axiomatization of Borda's rule. *Journal of Economic Theory*, 9(1):43–52, 1974. DOI: 10.1016/0022-0531(74)90073-8. 4

[You75] H. P. Young. Social choice scoring functions. *SIAM Journal on Applied Mathematics*, 28(4):824–838, 1975. DOI: 10.1137/0128067. 4

Authors' Biographies

DAVIDE GROSSI

Davide Grossi is a Lecturer (Assistant Professor) at the Computer Science Department of the University of Liverpool (UK). He holds a master's degree cum laude in Philosophy from the Scuola Normale Superiore of Pisa (Italy) and a Ph.D. in Computer Science from Utrecht University (the Netherlands). His Ph.D. thesis was nominated for the Christiaan Huygens Prijs 2009 of the Royal Netherlands Academy of Arts and Sciences. Prior to joining the University of Liverpool, he worked as a postdoctoral researcher at the Computer Science and Communications Department of the University of Luxembourg (on a personal grant sponsored by the National Research Fund of Luxembourg) and at the Institute for Logic, Language, and Computation of the University of Amsterdam (on a personal grant sponsored by the Netherlands Organisation for Scientific Research). He is author of over 40 peer-reviewed articles in international journals and conferences in philosophy, logic, artificial intelligence and multi-agent systems. He regularly serves as reviewer for top journals and conferences in his areas of expertise.

GABRIELLA PIGOZZI

Gabriella Pigozzi is a Maître de Conférences (Associate Professor) in Computer Science at Université Paris Dauphine (France) and a member of the LAMSADE Lab. After a Ph.D. in Philosophy from the University of Genova (Italy), she held postdoc positions at the Center for Junior Research Fellows of the University of Konstanz (Germany), at the Computer Science Department of King's College London (on a personal postdoctoral fellowship sponsored by the Economic and Social Research Council), and at the Computer Science and Communications Department of the University of Luxembourg. She obtained grants from the Engineering and Physical Sciences Research Council (UK) and the National Agency for Research (France). Her research focuses on judgment aggregation, computational social choice, argumentation theory, and normative multi-agent systems. She published over 40 peer-reviewed articles in international conferences and journals.

Index

INDEX

Printed in the United States
by Baker & Taylor Publisher Services